Doubling Farmers Income through Appropriate Agricultural Technologies and Extension Approaches

The Editors

Dr. Bhanu Prakash Mishra completed Bachelors from DDU, Gorakhpur University, Gorakhpur, M.Sc. (Ag.) from NDUAT, Kumarganj, Faizabad and PhD from Banaras Hindu University, Varanasi, U.P. He started his career as SMS (Agricultural Extension) at Krishi Vigyan Kendra, Khawzawl, Champhai district, Mizoram and served there for 2 years. After that he has joined as an Assistant Professor (Extension Education and Rural Sociology) at College of Horticulture and Forestry, Central Agricultural University, Pasighat, Arunachal Pradesh, India. Dr. Mishra has 10 years experience in teaching, research and extension education. He has written number of research papers, review papers, popular articles, bulletins, book chapters, *etc*. At present he is working as Associate Professor and Head, Department of Agricultural Extension, Banda University of Agriculture and Technology, Banda, U.P.

Dr. Brijesh Kumar Gupta, completed graduation degree, B.Sc. (Agriculture) from Purvanchal University, Jaunpur, U.P, M.Sc. (Agriculture) & PhD in Extension Education from Department of Extension Education, Institute of Agricultural Sciences, Banaras Hindu University, Varanasi, Uttar Pradesh. He started his career as Cane Supervisor in State Sugarcane Development Department at Maharajganj District of Uttar Pradesh and served ten years. After that, he has joined as an Assistant Professor in Department of Agricultural Extension at College of Agriculture, Banda University of Agriculture and Technology, Banda, Uttar Pradesh, India. Dr. Gupta has six years' experience in teaching, research and extension. He has also completed degree of Master of Business Administration (MBA) from Sikkim Manipal University, Sikkim. He has written a numbers of research papers, review papers, popular articles, bulletins, training manuals, book chapters *etc*. At present he is working as Assistant Professor in Department of Agricultural Extension at College of Agriculture, Banda University of Agriculture and Technology, Banda, Uttar Pradesh.

Dr. Dheeraj Mishra did his Ph.D. in Extension Education from Banaras Hindu University, Varanasi. He has more than 3 years of experience in teaching and research. He has published more than 14 research papers/book chapters and popular articles in Journals and Magazines of National and International repute. He is a life member of various academic societies. Presently Dr. Mishra is working as Assistant Professor in Department of Agricultural Extension, Banda University of Agriculture and Technology, Banda.

Dr. Pankaj Kumar Ojha did his M.Sc. (Ag.) in Extension Education from Banaras Hindu University, Varanasi and Ph.D. in Extension Education from Dr. Rajendra Prasad Central Agricultural University, Pusa. Besides these, he did Post graduate diploma in rural development from IGNOU and Post graduate diploma in Theatre communication from Banaras Hindu University, Varanasi. He has numbers of publications to his credit including books (two), research papers, book chapters, review articles, popular articles, *etc*. He has to credit a national award AIASA Gold medal for his literary contribution. Presently he is working as an Assistant Professor, Department of Agricultural Extension, Banda University of Agriculture and Technology, Banda, U.P.

Doubling Farmers Income through Appropriate Agricultural Technologies and Extension Approaches

– Editors –

Bhanu Prakash Mishra

Associate Professor and Head

Brijesh Kumar Gupta

Dheeraj Mishra

Pankaj Kumar Ojha

Assistant Professor

Department of Agricultural Extension,
Banda University of Agriculture and Technology, Banda

BIOTECH

2021

BIOTECH BOOKS®

ISBN 978-81-7622-455-0

Published by: **BIOTECH BOOKS**®
4762-63/23, Ansari Road, Darya Ganj,
New Delhi - 110 002
Phone: 09899295259, 011-43003222
E-mail: biotechbooks@yahoo.co.in
Website: www.biotechbooks.in

Typeset at: **Classic Computer Services**
Delhi - 110 035

Printed at: **Replika Press Pvt. Ltd.**

PRINTED IN INDIA

Foreword

Appropriate agricultural technologies are prerequisite for agricultural sustainability which includes sustainable production, economic viability and environmental quality. Agricultural technologies enabled with effective extension approaches are essential for empowering farmers, increasing farm productivity and farm income. In this context, it is necessary to disseminate appropriate and location specific agricultural technologies suitable for various categories of farmers as well as resources available at farmer's doorstep by using different extension approaches.

I congratulate the Department of Agricultural Extension, College of Agriculture for organizing a 21 Days Winter School on 'Appropriate Agricultural Technologies and Extension Approaches for Doubling Farmers' Income' in collaboration with National Institute of Agricultural Extension Management (MANAGE), Hyderabad.

I appreciate that the Department has taken up the initiative to compile all the lectures delivered during winter school and to publish them in the form of edited book. This book covers technologies of various aspects such as Crop sector (field crops and horticulture), Livestock sector (dairy, poultry, piggery, small ruminants), Farm linked activities (mushroom, beekeeping, sericulture), Post production activities (off-farm enterprises) *etc*. It also revisits the resources and capacities needed in extension, in the light of modern day technology, including ICT, physical connectivity and the increased presence of NGOs and private sector in providing extension services to farmers. Further, I also appreciate the efforts being made by the editorial team for shaping this worthy publication.

I hope this book would be useful for academicians, researchers, extension professionals and students for their academic endeavor.

U.S. Gautam

Preface

Agriculture is the primary source of livelihood for majority of India's population. It engages 54.6 per cent of population and contributes 17.4 per cent in GDP. Since independence, we have achieved a significant increase in overall food grain production through execution of agricultural research and extension programmes. But, it is also evident that imbalanced development has increased the inequality among various sections of rural societies.

In the present time, the most challenging issue is to shift from production to income. The focus of the efforts needs to be reoriented so that it may lead to socio-economic upliftment and long-term food and nutritional security of the rural population. With this concern, Government of India laid forward the visionary idea of doubling farmers' income by 2022. It is hoped, that the answer to agrarian challenges and realization of the aim of farmers' welfare lies in higher and steady incomes. Agricultural household activities comprise of Crop Sector (field crops and horticulture), Livestock Sector (dairy, poultry, piggery, small ruminants), Farm Linked Activities (mushroom, beekeeping, sericulture), Post Production Activities (off-farm enterprises, secondary agriculture). The appropriate agricultural technologies and effective extension strategies are two wheels of development cart leading to sustainability of various farm or off-farm enterprises.

Keeping this in view, a MANAGE Sponsored 21 Days Winter School on 'Appropriate Agricultural Technologies and Extension Approaches for Doubling Farmers' Income' was organized by Department of Agricultural Extension, College of Agriculture, Banda University of Agriculture and Technology, Banda, Uttar Pradesh during 5th to 25th February 2019 in university campus.

This book is a compilation of the lectures presented by the eminent academicians and researchers during the winter school. We hope, this compilation will really be very fruitful for all the stakeholders including Subject Matter Specialists, Scientists, Students and Progressive Farmers in improving their knowledge about recent trends, technologies and approaches in agriculture and allied sectors.

Editors are pleased to express their gratitude to all the contributors. We also acknowledge the guidance and motivation received from Dr. U.S. Gautam, Hon'ble Vice Chancellor, Banda University of Agriculture and Technology, Banda, U.P. We also express our sincere gratitude to Smt. V. Usha Rani, DG, MANAGE and Dr. R. Saravanan, Director (Agril. Extension), MANAGE, Hyderabad for extending all sorts of help and support from organizing winter school to publication of this book.

The editors will be grateful to the readers, if they point out the error and send suggestions and constructive criticism for future revision and improvements of the book.

Bhanu Prakash Mishra

Brijesh Kumar Gupta

Dheeraj Mishra

Pankaj Kumar Ojha

Contents

Chapter 1

Fodder Production Technologies for Enhancing the Farm Income

Anoop Kumar Dixit[1] *and Jagannath Pathak*[2]

[1]*ICAR-Indian Grassland and Fodder Research Institute, Jhansi – 284003, U.P.*

[2]*Department of Soil Science and Agricultural Chemistry, CoA, BUAT, Banda – 210 001, U.P.*

Traditionally, Indian agriculture has been a mixed farming system where crop production and livestock remained the major functional and complimentary components besides providing livelihood for more than 60 per cent population in rural areas. The livestock not only provide products for human diets but also serve as energy source for draught power in different agricultural operations of small and marginal farmers. In arid and semi-arid regions, livestock is the only source of livelihood, particularly when agriculture fails to withstand the drought. The country rears >10 per cent of the world's livestock population on 2.3 per cent of geographical area (DAHDF, 2014). As per the 19th livestock census (2012), the country is rearing 512 million livestock including 191 million cattle and 108 million buffaloes with total milk production of 165 mt in the year 2016-17. In developing countries like India, it is predicted that meat and milk consumption will grow at 2.8 and 3.3 per cent per annum, respectively. The low productivity of Indian livestock is mainly attributed to poor fodder and feed resources.

Forage Scenario

Forages are plants which are grown and to be utilized by grazing or harvesting as a whole crop whereas Fodder crops are plant species that are cultivated and harvested as green for feeding the animals (cut green and fed fresh), making silage and hay. The requirement for fodder is mainly governed by the population and species of livestock, management levels and climatic conditions. As per estimates of IGFRI, the availability of green fodder from cultivated areas and pastures is estimated to be 590 mt against the demand of 851 mt in the country by 2020. Similarly, in case of dry fodder, estimated total availability will be 467 mt (from dry residues and by-products of grain, commercial and pulse crops and dry grass from grazing lands and forests) against the requirement of 530 mt (IGFRI, 2013).

The total area under cultivated fodders is 8.3 million ha on individual crop basis. Sorghum amongst the *kharif* crops (2.6 million ha) and berseem (*Egyptian clover*) amongst the *rabi* crops (1.9 million ha) occupy about 54 per cent of the total cultivated fodder cropped area (ICAR, 2009). The area under cultivated fodder crops is static since last two decades. The scope for further increase in area under forages seems meager due to demographic pressure of human population for food crops. It can be increased by growing fodder and grasses on salt affected soil. The tailor made information will be helpful for informed decision making by the farmers, students, and development agencies, forage research and development personals. Thus, it will be helpful in meeting the overall objective of providing adequate green and dry forage for balanced feeding to meet the demand of meat, milk and other products.

Regional Imbalances in Fodder Availability

The Green Fodder Availability

i. In Western Himalayan (1), Upper Gangetic Plains (5) and Eastern Plateau and Hilly Zones (7) is more than 60 per cent of the actual requirement

ii. In Trans Gangetic Plains (6) is between 40 - 60 per cent of the requirement

iii. In the remaining zones, the figure is below 40 per cent.

Dry Fodder Availability

i. In the Eastern Himalayan (2), Middle Gangetic Plains (4), Upper Gangetic Plains (5), East Coast Plains and Hilly Zones (11) >60 per cent of actual.

ii. In Trans Gangetic Plains (6), Eastern Plateau and Hills (7) and Central Plateau and Hills (8) between 40-60 per cent,

iii. Less than 40 per cent in the remaining zones of the country.

Salient Features of Forage Crops

i. Shorter growth period,

ii. Grown in closer spacing with high seed rate,

iii. Dense stand to smother weeds and prevent soil erosion,

iv. Improve soil health through addition of higher amounts of organic residues in the soil,

v. Crop duration can be adjusted and risk due to aberrant weather conditions minimized,

vi. High persistency and regeneration capacity reduce the need for frequent sowing and tillage,

vii. Wider adaptability with capacity to grow under stress conditions,

viii. High nutrient and water requirement under intensive cropping.

Production and Management of Fodder Crops

Forage crops can be categorized in to 2 groups *viz.*

i. ***Cereal fodders*** provide energy-rich herbage to livestock. These have wider adaptability and variability in terms of growth, regeneration potential, yield and quality of herbage. *viz.*, Sorghum, Maize, Pearlmillet, Oats, *etc.*

ii. ***Leguminous fodders*** have special significance because of high herbage protein and partial independence from soil for their nitrogen needs. *viz.*, Cowpea, Berseem, Lucerne, *etc.*

The detailed package and practice of fodder cultivation (Kumar, *et al.*, 2012) and promising production technologies are discussed as under:

1. Choice of Suitable Forage Crops and their Variety

Fodder crops are the plant species that are cultivated and harvested for feeding the animals in the form of forage (cut green and fed fresh), silage (preserved under anaerobic condition) and hay (dehydrated green fodder).

i. *Leguminous fodders:* The leguminous fodders have special significance because of high herbage protein and partial independence from soil for their nitrogen needs.

ii. *Non-leguminous fodders: Non-leguminous* fodders provide energy-rich herbage to livestock.

Identification of forage crops and their variety according to soil type, area and situation is required for increased crop yield and reduced yield variability, increased resilience against climate change, particularly increases in climate variability (prolonged periods of drought, seasonal shifts in rainfall, *etc.*) and increased soil carbon storage. Menhi Lal 1993 suggested following desirable characteristics in a forage variety.

i. Growth rhythm with early bulkiness and matching physiology for higher rate of dry matter accumulation and longer leaf area duration.

ii. Compatibility with associated crops in mixed/intercropping systems.

iii. High palatability, greater digestibility, better nutritive value and free from anti-quality constituents.

iv. Quick regeneration potential, flexibility to defoliation management and greater persistency.

v. Ability to withstand aberrant weather and adverse soil conditions such as drought, water logging, saline-sodic and acidic soils.

vi. Multiple resistances to diseases and insect pests.

vii. High nodulating and non-structural carbohydrates reserve abilities in forage legumes for subsequent regeneration following frequent cuttings.

viii. Desirable characteristics for making good quality silage and hay.
ix. Suitability for fitting in the prevalent cropping systems.
x. Acceptable seed production ability and/or dual-purpose types.

The important high yielding cultivated forage crop varieties under different situations are enlisted as under:

Crop	*Variety*	*Area of Cultivation*	*Green Fodder Yield (t/ha)*
Sorghum	Single cut		
	PC-6, 9, 23 HC- 171, 260	Whole country	35-50
	HC-136, Raj. Chari- 1 and 2	Whole country	37-50
	Multi cut		
	PCH-106	North India	60-90
	SSG-988, 898, 855	Whole country	75-105
Maize	African Tall	Entire country	55-80
	Vijay, Moti and Jawahar composite	Entire country	35-47
	L-72 and 74	Entire country	40-42.5
	J-1006, Pratap Makka Chari-6	North India	45-55
	VL-54	Hilly area	30-45
Bajra	Raj Bajra	Entire country	30-45
	Giant bajra, multi cut	Entire country	55-170
Cowpea	IGFRI-450	North India	30-40
	UPC-5287	North India	35-40
	IFC-8503, EC-4216	North, west and central India	30-40
	UPC-5286	Whole country	30-45
	IFC-8401	Whole country	30-45
	UPC-4200	North-East region	35-45
Guar	BG-1	Entire guar growing tract	22-35
	BG-2	Entire guar growing tract	20-38
Berseem	Mescavi	All berseem growing region	70-90
	Wardan (S-99-1)	All berseem growing region	80-110
	JB -1	Central and North-Western zone	85
	UPB-110	Central and South zone	55-70
	JB-2 and JB-3	Central zone	70-95
	BB-2	Central zone	90-100
Lucerne	Sirsa-8	Northern zone	80-85
	Anand-2	Gujarat, Rajasthan and MP	80-100
	RL 88	Entire growing tract	75-100
	T-9	Northern zone	80-95
Oat	HFO-114 and Kent		45-50
	OS-6 and OS-7		40-55

Crop	*Variety*	*Area of Cultivation*	*Green Fodder Yield (t/ha)*
	UPO-94 and UPO-212		45-60
	Bundel Jai-851		40-50
	Bundel Jai-2004		45-50
	Bundel Jai-822		40-45
Barley	Azad and DL-36		15-20 (1.5-3.5 grain

Cropping Systems

Cropping systems should be selected in such a way that it should also help in stability and sustainability of soil fertility in long run, minimize harbouring of insect-pest and diseases.

i. Guinea grass/NB hybrid + (Cowpea – Berseem/lucerne)

ii. Guinea grass/NB hybrid + (Rice bean/Cowpea – Lucerne/Berseem)

iii. Other cropping systems: Sorghum - Berseem, Sorghum –Oat + Lathyrus, Sorghum – Oat + Berseem, Sorghum – Oat (multicut), Sorghum + Cowpea – Oat, Sorghum + Cowpea –Berseem + Mustard, Sorghum – Lathyrus + mustard, Sorghum (grain) – Lucerne - Maize (fodder), Maize + Cowpea – Oats – Maize + Cowpea, Cowpea + Maize (fodder) – Lucerne, Maize/ rice/sorghum/green gram/soybean – Lucerne.

2. Soil and its Preparation

Crop	*Soil Type*	*Land Preparation*
Sorghum	Clay loam	Soil opened with soil inversion plough followed by 1-2 operations by cultivator and planking
Maize	Well drained loamy soil	-do-
Pearlmillet	Well drained sandy loam soil	-do-
Cowpea	Well drained loam to clay loam soil	-do-
Berseem	Well drained clay to clay loam soils. It can be grown in saline-sodic soils	Soil opened with soil inversion plough followed by 2-3 operations by cultivator and planking
Lucerne	Well drained fertile soils with neutral pH are ideal	-do-
Lathyrus	Sandy loam to clay loam soils	Soil opened with soil inversion plough followed by 1-2 operations by cultivator and planking
Oats	Performs well on loam to clay loam soil with adequate drainage	Soil opened with soil inversion plough followed by 2-3 operations by cultivator and planking
Barley	Sandy loam to clay loam soils	Soil opened with soil inversion plough followed by 1-2 operations by cultivator and planking

Seed and Sowing

Crop	*Sowing Time*	*Seed Rate (kg/ha)*	*Spacing and Depth*
Sorghum	March-July	25-35	30x10 cm, 3 cm
Maize	April - July	40-50	30x15 cm, 3-4 cm
Pearlmillet	March-July	12-15	25x10 cm, 2 cm
Cowpea	March-July	20-25	30x 10 cm, 3 cm
Cluster bean	April - July	25-40	30-45 cm R to R
Berseem	October	25	25 cm R to R or broadcasting
Lucerne	Mid of October	12-15	25 cm R to R, 2 cm
Lathyrus	October to November	50	30-35 cm R to R, 4 cm
Oat	Mid October to last November	80-100	20-25 cm R to R for low tillering and 25-30 cm R to R for high tillering varieties, 4 cm depth
Barley	October to November	80-100	20-25 cm R to R, 4 cm
Berseem	October	25	25 cm R to R or broadcasting
Lucerne	Mid of October	12-15	25 cm R to R, 2 cm
Lathyrus	October to November	50	30-35 cm R to R, 4 cm
NB Hybrid	March to August	20000 to 35000 rooted slips	100x50 to 75x50 cm
Guinea grass	March to August	20000 to 40000 rooted slips	100x50 to 50x50 cm

3. Nutrient Management

Mineral Nutrition

Since centuries, it is known that roots of terrestrial plants obtain nourishment from the soil. During the first half of the nineteenth century, it was found that plants need certain chemical elements referred to as essential elements and that elements are absorbed by roots principally as inorganic ions. The term mineral nutrient is generally used to refer as inorganic ion obtained from the soil and required for plant growth. The process of absorption, translocation and assimilation of nutrients by the plants is known as mineral nutrition.

Essential Elements

Plant needs 17 elements for their growth and completion of life cycle. They are: carbon, hydrogen, oxygen, nitrogen, phosphorus, potassium, calcium, magnesium, Sulphur, iron, manganese, zinc copper, boron, molybdenum, chlorine and nickel. In addition, four more elements *viz.*, sodium, cobalt, vanadium, and silicon are absorbed by some plants for special purposes. All these elements are not required by all plants. Among these, all carbon atoms and most of the oxygen atoms are derived from the carbon dioxide which is assimilated principally in photosynthesis.

Classification of Essential Elements

The essential elements can be classified based on the amount required, their mobility in the plant and soil, their chemical nature and their functions inside the plant.

Amount of Nutrients

Depending on the quantity of nutrients present in plants, they can be classified into three:

Basic Nutrients

The basic nutrients *viz.* carbon, hydrogen and oxygen, constitute 96 per cent of total dry matter of plants. Among them, carbon and oxygen constitute 45 per cent each. The total dry matter produced by rice crop in one season is about 12 t ha^{-1}.

Macronutrients

The nutrients required in large quantities are known as macronutrients. They are N, P, K, Ca, Mg, and S among these, N, P, and K are called primary nutrients and Ca, Mg and S are known as secondary nutrients. The later are known as secondary nutrients as they are inadvertently applied to the soils when N, P, and K fertilizers which contain these nutrients are used.

Micro-nutrients

The elements which are required in small quantities are known as micro-nutrients or trace elements. They are Fe, Mn, Zn, Cu, B, Mo, and Cl. These elements are very efficient and minute quantities produce optimum effects. On the other hand, even a slight deficiency or excess is harmful to the plants.

Chemical Nature

The nutrients can be classified into cations and anions and metals and non-metals based on their chemical nature

Cations	K, Ca, Mg, Fe, Mn, Zn,Cu
Anions	NO_3, H_2PO_4, SO_4
Metals	K, Ca, Mg, Fe, Mn, Zn,Cu
Non- metals	N, P, S, B, Mo, Cl

Functions in Plants

Based on the functions, nutrients are grouped into four:

1. Elements that provide basic structure to the plant – C, H and O.
2. Elements useful in energy storage, transfer and bonding – N, S, and P.
3. Elements necessary for charge balance – K, Ca, and Mg. these act as regulators and carriers.
4. Elements involved in enzyme activation and electron transport – Fe, Mn, Zn, Cu, B, Mo, and Cl. These elements are catalysers and activators.

Factors Influencing Nutrient Availability

1. Natural supply of nutrients in the soil which is closely tied up to parent material of that soil and vegetation under which it is develop.
2. Soil pH as it affects nutrient release.
3. Relative activity of micro-organism which plays a vital role in nutrient release and may in the case of mycorrhiza directly function in nutrient uptake.
4. Fertility addition in the form of commercial fertilizers, animal manure and green manure.
5. Soil temperature, moisture and aeration.

Integrated Nutrient Management

Indian soils are cultivated with negative nutrient balance of 12-14 Mt per year and this negative balance will continue to exist even after exploring full potential of fertilizer industry. Under this situation, it is really very hard to maintain soil health and productivity. Adequate plant nutrient supply holds the key for improving the food grain production and sustaining the soil health. The supply of nutrients in the form of organic manures helps in retaining more moisture, thereby enhancing the water and nutrient use efficiencies. Integrated use of inorganic and organic sources of plant nutrients has a tremendous potential not only in sustaining agricultural productivity and soil health but also in substituting a part of fertilizer requirement by organics for different crops and cropping systems. In order to improve the supplies of all the essential plant nutrients and also to maintain it good soil health, it is necessary to integrate all the nutrient supply sources. Adoption of integrated plant nutrient supply and management strategies for enhancing soil quality, input use efficiency and crop productivity is extremely important for food and nutritional security in Indian agriculture.

Forages are heavy feeder, on an average 1tone of dry perennial grass remove 9.4 kg N, 1.45 kg P, 14.2 kg K, 4.61 kg Ca, 2.65 kg Mg, and 1.85 kg S. Response of per kg N application in annual forage varied 161-431 kg in green fodder and 34-85 kg in dry forage. Response of P fertilization varied 53-508 kg green forage and 11-69 kg dry forage yield per kg of P_2O_5 application.

NUE Enhancing Technique

- Split application of nitrogen.
- After every cut 25 kg N/ha should be applied in cultivated perennial grasses
- Proper placement of P, K and Micro-nutrient
- Crop rotation (cereal- legume): it reduces about 25 per cent N requirement
- Green manuring/brown manuring

Nutrient Management in different Fodder Crops

Crop	*FYM (t/ha)*	*Nitrogen*	*Phosphorus*	*Potassium*	*Nitrogen Top Dressing*
Sorghum (S)	25	30 kg	40 kg	20 kg	30 kg N/ha 30 DAS
Sorghum (M)	25	30 kg	40 kg	20 kg	30 kg N/ha 30 DAS, 30 kg N/ha just after cutting
Maize	25	30 kg	40 kg	20 kg	30 kg N/ha 30 DAS
Pearlmillet	25	25 kg	20 kg	12.5 kg	25 kg N/ha 30 DAS
Cowpea	25	25 kg	40 kg	20 kg	-
Cluster bean		15 kg	90 kg	20 kg	
Berseem	-	20	80-90	30-40	-
Lucerne	-	20	60-75	40	-
Lathyrus	-	20	40	-	-
Oat (one cut)	15	60	40	-	20 kg N after Ist irrigation
Oat (two cut)	15	80	40	-	40 kg N after Ist cut
Oat (multi cut)	15	100-120	40	40	40 kg N after each cut
Barley	-	45	30	20	15 kg N after Ist irrigation
NB Hybrid	15-25	40	50	50	30 kg N after each cut
Guinea grass	15-25	50	50	50	30 kg N after each cut

4. Irrigation Management

During monsoon season, generally irrigation is not required but if dry spell prevails more than 15 days then irrigation is needed for successful completion of life cycle. Whereas during rabi an summer season frequent irrigations are required depending on soil type, climate and crop. Proper irrigation scheduling- 25 per cent (in fodder crops) and 50 per cent (cereals) of depletion of available soil moisture. Promote drip (85 per cent IUE) and sprinkler irrigation- (70 per cent IUE) to save irrigation water. Irrigation should be applied as per critical stages of the crops. Adoption of FIRB helps in saving of about 20-25 per cent of irrigation water in deferent crops:

Crop	*Irrigation No./Scheduling*
Berseem	October to February: 14-16 days interval (Clay and clay loam soil), 12-14 days interval (Loamy soil) October to February: 10-12 days interval (Clay and clay loan soil), 8-10 days interval (Loamy soil)
Lucerne	15-20/At early stage:7-10 days interval. Later on: 25-30 days interval During summer: 15-20 days interval.
Lathyrus	On conserved soil moisture
Oat	First: 20-25 DAS, 4-5 (Single cut) and 7-8 (Double cut)

5. Weed Management

In forages chemical weed management should be avoided and to suppress weed population Deep summer tillage, clean cultivation, Manual weeding, Practice of weed mulch, Under perennial grasses at establishment stages- intercropping of BNH/Guinea + cowpea should be practiced.

Herbicide Based Weed Management Strategies for Fodder Crops

Crop	*Strategies*
Sorghum	1. Broadcast/cross sowing of cowpea (25 kg/ha) 2. Pre-emergence and post emergence (at initial growth of forage) application of Atrazine (0.50-0.75 kg/ha), pre-emergence application of Pendimethalin (1.0 kg/ha).
Maize	1. Intercropping with cowpea (25 kg/ha) 2. Pre-emergence application of Atrazine (0.75 kg/ha), Pendimethalin (1.0 kg/ha) with 600 litre of water after sowing on same day or positively by next day before the emergence of seedlings of the crop.
Pearlmillet	1. Pre-emergence application of Atrazine (0.50 kg/ha) with 600 litre of water after sowing on same day or positively by next day before the emergence of seedlings of the crop.
Oat	1. Pre-emergence application of Linuran (0.5 kg/ha) 2. Post-emergence application of 2,4- D sodium salt (0.75 kg/ha)
Cowpea	1. Pre-plant incorporation of Trifluralin (0.75 kg with 600 litre of water/ha) herbicide.
Guar	1. Pre-plant incorporation of Nitralin (0.75 kg/ha) herbicide.
Berseem	1. Dipping the seed in 5-10 per cent solution of table salt for five minutes, light weight weed seeds float while heavy weight forage seeds will settle down at the bottom. 2. Use stale seed bed technique: Pre-sown irrigation followed by spraying of Paraquat herbicide for newly germinated weeds. 3. Pre-plant incorporation of EPTC (1.5 kg/ha) herbicide. 4. Pre-emergence application of Butachlor (2.0 kg/ha), Pendimethalin (0.75 kg/ha) 5. Post emergence (3-4 trifoliate leaf stage) application of MCPB (0.50 kg/ha) herbicide for control of broad leaf weeds. Note: EPTC should not be used in fields where a Trizine herbicide was applied in the previous season.

Crop	Strategies
Lucerne	1. Dipping the seed in 5-10 per cent solution of table salt for five minutes, light weight weed seeds float while heavy weight forage seeds will settle down at the bottom. 2. Use stale seed bed technique: Pre-sown irrigation followed by spraying of Paraquat herbicide for newly germinated weeds. 3. Pre-plant incorporation of Fluchloralin (1.0 kg/ha), Benefin (0.75-1.5 kg/ha herbicide. 4. Pre-emergence application of Pendimethalin (0.5-1.0 kg/ha) and Oxadiozon (0.5 kg/ha) herbicide. 5. Post emergence (3-4 trifoliate leaf stage) application of 2, 4-DB (0.50-0.75 kg/ha) herbicide for control of broad leaf weeds.

6. Harvesting Management

Harvesting management of forage crops is very crucial for getting quality green fodder. Harvesting schedule of different fodder crops is summarized below:

Crop	Stage of Harvest	Time of Harvesting
Sorghum (Single cut)	After flowering and up to milking stage	75-80 days after sowing
Sorghum (Multi cut)	-	2 months after sowing and subsequent cuts 35-40 days after the previous cut
Maize	Cob formation to milk stage	60-70 days after sowing
Bajra (Single cut)	Boot leaf stage to early flowering	60-75 days after sowing
Bajra (Multi cut)	-	Ist cut 55-60 days after sowing and subsequent cuts 35-40 days after the previous cut
Cowpea	At 50 per cent flowering	50-60 days after sowing As mixed crop at the time of companion crop
Guar	-	55 days after sowing
Berseem	-	1st cut: 50 – 55 days after sowing and subsequent 25-30 days (5-6 cuts)
Lucerne	-	1st cut: 65-80 days after sowing and subsequent 25-30 days (7-8 cuts)
Lathyrus	50 per cent flowering	65-70 days after sowing
Oat (single cut)	50 per cent flowering stage	60 days after sowing
Oat (multi cut)	-	Double cut 1st: 60 DAS and II cut at 50 per cent flowering Multi cut 1st: 60 DAS, II cut 45 days and III cut at 50 per cent flowering
Berseem	-	1st cut: 50 – 55 days after sowing and subsequent 25-30 days (5-6 cuts)
NB hybrid and Guinea grass	-	Ist 75 days after planting, subsequent at 30 days interval

Technology- I: Round the Year Fodder Production System (Irrigated Situation)

Dimension: The system comprises of raising seasonal fodder crops, inter-planted with perennial grasses (hybrid napier/guinea grass). Hybrid Napier based cropping system (Hybrid Napier + (Cowpea - Berseem + Mustard) has green fodder production potential of 273 t/ha and dry fodder potential of 44.3 t/ha per year under assured water supply (Water requirement 1090 mm).

Target area: The system is suitable for large scale dairy farmers in Peri-urban and Milkshed areas of whole India except tropical region.

Impact: Technology can supply round the year good quality fodder (Cereal: legume, 67:33) which can sustain 8-10 ACU per ha (1 ACU= 350 kg body weight). The technology is being demonstrated in farmer's field in Bundelkhand.

Technology-II: Round the Year Fodder Production System (Rainfed Situation)

Dimension: The Institute has developed Subabul + Trispecific hybrid - Fodder sorghum + Pigeonpea based fodder production system for rainfed conditions. In this system, the *Pennisetum* Trispecific Hybrid (TSH) is planted in paired rows at 0.75 m x 0.5 m spacing. Subabul is planted at 50 cm plant to plant spacing in between pairs of TSH. The 3 m space between such two alleys is utilized for fodder sorghum + pigeonpea cropping system in 2:1 ratio at 30cm.

Target area: Small and medium farmers of semi arid region (up to 500 mm rainfall)

Impact: This system has green fodder production potential of 53.3 t/ha and 13.28t dry fodder besides producing 0.4 t/ha pigeon pea grain and 0.8 t/ha dry sticks having fuel value. Technology can sustain 2-3 ACUs with quality fodder. It also provides nutrient rich pulse for human consumption and fuel wood. It is also capable of prolonging the fodder availability up to May (dry period).

Technology-III: Fodder on Field Boundary/Bunds/Channels: Noncompetitive Land Use

Dimension: Among different perennial cultivated grasses, Napier - bajra hybrid is most suitable for bunds of irrigated areas and Tri-specific hybrid (TSH), guinea grass, anjan grass and nandi grass are suitable under rainfed conditions.

Target area: All categories of farmers' especially marginal and small farmers in different agro climatic regions of India

Impact: Fodder yield recorded from different farm demonstrations indicated 1.75 to 2.50 kg green fodder per meter per cut and on an average in 4 cuttings 7.0- 11.0 q green fodders per 100 metre bund length is possible in a year. Besides additional farm productivity, it also works as a guard crop for main crop, reduces runoff loss of water and controls soil erosion. Now this technology is being disseminated in different rural parts of country.

References

DAHDF. 2014. 19th Livestock Census – 2012. All India Report. Department of Animal Husbandry, Dairying and Fisheries, Ministry of Agriculture, Government of India, New Delhi. Retrieved from http://da hd.nic.in/sites/default/filess/ Livestock per cent 20 per cent 205.pdf

ICAR. 2009. Forage Crops and Grasses. *In*: Handbook of Agriculture. Indian Council of Agricultural Research, New Delhi.

ICAR. 2010. Degraded and Wastelands of India: Status and Spatial Distribution. Indian Council of Agricultural Research, New Delhi.

IGFRI. 2013. Vision 2050. Indian Grassland and Fodder Research Institute, Jhansi, Uttar Pradesh, India.

Kumar Sunil, Agrawal R.K., Dixit A.K., Rai Arvind K., Singh J.B. and Rai S.K. 2012. Forage Production Technology for Arable Lands. ICAR-Indian Grassland and Fodder Research Institute, Jhansi-284003.

Chapter **2**

Soil Quality Assessment: Concept and Method for Identification of Soil Quality Indicators and Computing Soil Quality Index

S.R. Singh[1], Poonam Yadav[1], A.K. Chaube[2] and M.K. Tripathi[1]

[1]ICAR-Indian Institute of Sugarcane Research, Lucknow
[2]Banda University of Agriculture and Technology, Banda

Agricultural sustainability has become a vital issue in developing countries, including India. Over-exploitation of soils owing to intensive cultivation and imbalance fertilization over the decades has resulted to steadily declining factor productivity has been noticed in long term experiments in Asia (Manna *et al.*, 2005). The traditional rice-wheat cropping system exhibited sign of yield stagnation and decreasing factor productivity during last two decades due to declining soil quality, depleting ground water, soil degradation, rising problems of salinity and sodicity, declining soil organic carbon, multiple micronutrient deficiency and large-scale burning of rice straw in rice-wheat cropping system (Bhattacharyya *et al.*, 2015; Bhatt *et al.*, 2016; Yadav *et al.*, 2016). The declining trend in rice productivity was first noticed in long-term experiments notably in rice-wheat cropping system of India. The wetland rice cultivation due to puddling and keep area flooded from entire duration (planting to grain filling) of rice crop under rice-wheat rotation may destroys soil structure and creates poor physical condition which led to decreased wheat yield by limiting root growth and its distribution. Besides declining soil fertility, lower wheat yields is the main causes of less profitable in a short turnover period between rice harvest and delayed wheat sowing. The delayed rice transplanting and harvesting caused by persisting high soil moisture content after the rice harvest liable to increased input cost involved in conventional tillage practices under traditional rice-wheat cropping system.

Day by day, soil health is declining at an alarming rate due to application of imbalance nutrition and over exploitation of natural resources caused nutrient deficiencies not only in macro-nutrient levels but also micro-nutrients in the soils of this region. Kukal and Aggarwal (2003a) stressed on intensity of puddling operations which reduces losses where soils are highly permeable and coarse textured. However, its negative impacts on structural degradation notably in upland crops are real concern. The repeated puddling of coarse and medium textured soils has led to the sub-surface compaction which is detrimental for the upland crops specially wheat. Intensive cultivation may be the reason for breaking large aggregates which further leads poor contact with seed, thereby declining potential yield. Thus, puddling operations are water, capital and energy intensive which finally deteriorates the soil structure. Singh (2000) reported that 3 percents soils of Haryana had low P content during 1980 but it jumped to 73 per cent by 1995 whereas N deficient areas increased only in extent of non-significant (89 to 91 per cent) during that period. Soils with higher K values had come down from 91 per cent during 1980 to 61 per cent in 1995. Due to intensive rice-wheat cropping system and imbalance fertilization disturbed the nutrient balance in upper zone of soil (Gill, 1992). Conventional practice of crop residue burning particularly, rice straw cause serious consequences such as loss of nutrients and emission of green house gases. Farmers has to put more input specially applied higher doses of fertilizers to achieve higher potential yield resulting higher cost of cultivation in the short term but deteriorated soil quality and decreases crop productivity in long term.

Before mid sixties, macro-nutrients supplied through commercially available mineral fertilizers, however, with time, the soil quality started to decline and micro-nutrients deficiencies surfaced where rice-wheat sequence extensively carried out (Biswas and Tewatia, 1991). Many workers reported that yield advantage was recorded due to sulphur application in rice-wheat system in Indo-Gagetic plains of India (Tiwari *et al.,*, 1989; Katyal, 2003). Hence, quantified recommendation of sulphur in textured soil is needed for better crop productivity and soil quality. Wheat grown after rice suffered from manganese deficiency notably in coarse textured soils of Punjab. The boron deficiency was observed in West Bengal which declined wheat yield under rice-wheat cropping system. Soil application of borax corrected the deficiency and enhanced wheat yield substantially. The selenium toxicity's is also an emerging problem coming out of intensive rice-wheat cropping system of this region but maize-wheat system had no symptoms of selenium toxicity. The appearance of iron and zinc deficiency in rice whereas manganese and selenium toxicity in wheat are the new hindrance towards sustainability of rice-wheat cropping system. It is very difficult to farmers that can neither recognized them at exact time nor ameliorate them by applying their correct doses either through spray or through broadcasting.

Decline crop response due to fertilization is the major causes for the declining land productivity. The annual consumption of the NPK fertilizers was 28 Mt in

India while actual use stands at only 18 Mt of NPK, thus creating negative balance of about 10 Mt of primary nutrients. Yield decline in rice-wheat was greater when N-fertilizers applied was alone at the rate of 120 kg N ha^{-1} while as response of P and K fertilizers to crop yield was started after 10 and 5 years, respectively. The main causes behind low crop response to N-fertilizers are its low use efficiency (only 30-40 per cent of applied N-fertilizers in rice) owing to surface runoff, leaching, volatilization and denitrification. Some other identified causes for low crop response and declining crop productivity are intensive cropping system, injudicious use of mineral fertilizers, faulty methods and timing of fertilizers application, inadequate use of organic manures, excessive use of N-fertilizers, antagonistic reaction between some plant nutrients, soil degradation and lack of quality soil testing facilities.

The concepts of soil quality, soil health, and soil quality/health assessment are highly contentious within the soil science community, because many believe these terms have generalized and oversimplified the collective knowledge and wisdom developed through several centuries of intensive, in depth, global studies of soil resources (Letey *et al.*, 2003; Sojka *et al.*, 2003). A common theme is that soil quality/health assessments are impossible and meaningless because of the complexity of soil resources. They suggest research and education should be focused on developing quality soil management practices rather than on soil quality or soil health. Proponents of soil quality argue that soil scientists have long recognized the many unique and important properties and processes provided by fragile soil resources, outside the agricultural community, soils remain largely an under-valued resource (Karlen *et al.*, 2003). The assessments are viewed as tools intended to alert users, in a manner analogous to a "consumer price index," that soil resource problems have or may be occurring.

Why is Soil Quality Assessment Required?

Periodic assessment is required to evaluate the condition of soil resources at field, farm, state and nation levels because humankind has neglected its soil resources regularly in the dominant society and culture (Hillel, 1991). Despite this well-documented history, degradation of soil resources existed in the earth's is still among the most serious and widespread threat to humankind. With very little effort, we can find gullies cutting large fields into small parcels, road ditches that have to be cleaned out, silt-laden streams, lakes being choked by sediment, and windstorms with blowing soil darkening western skies. These are such visible signs of soil degradation which may not surprise tolerable soil loss and defined as the maximum amount of erosion at which the quality of a soil as a medium for plant growth can be maintained, became the primary tool used to assess sustainability of soil resources. Soils can also be degraded by excess water, sodicity, salinity, compaction, heavy metals, acidification, and loss of nutrients and organic matter. Since these degraded conditions exist on millions hectares of lands not only in Indo-Gangetic plains of India but all over India or worldwide (Oldeman, 1994). It is essential that more robust assessment tools be developed for assessing soil quality regularly. Current efforts to define soil quality/health and develop multi-factor assessment protocols can be traced to publications from the 1970s (Alexander, 1971; Warkentin and Fletcher, 1977). This coincided with increased emphasis on

"Sustainable Agriculture" during the mid to late 1980s (*e.g.* NRC, 1989) that brought public attention to the increasing degradation of soil resources and the implications for environmental health. Many questions were arises regarding the sustainability of current soil and crop management decisions (Pesek, 1994), discussion and interest in the concepts of soil quality and soil health spread worldwide (Karlen *et al.*, 1997; 2001). Several ideas for assessment evolved following publication of quantitative formula for assessing soil quality and efforts to relate changes in various indicators to soil management practices (Karlen *et al.*, 1994a, b). Interest in soil quality among natural resource conservationists, scientists, farmers and policymakers increased even more after the U.S. National Academy of Sciences published the book entitled *Soil and Water Quality: An Agenda for Agriculture* (NRC, 1993). This report revealed that more holistic research was needed to ensure soil resources were sustained, water quality was protected, and money invested in conservation was well spent. Among the responses to those challenges were the reorganization of the USDA-Soil Conservation Service (SCS) to the USDA-Natural Resources Conservation Service (NRCS), creation of several Institutes including the USDA-Soil Quality Institute, development of user-oriented soil quality scorecards and test kits (Romig *et al.*, 1996), and several symposia (Doran and Jones, 1996) that defined soil quality, identified critical soil functions, and proposed applicable assessment methods (Doran and Parkin, 1994).

Doran and Parkin (1994) defined soil quality as the "capacity of the soil to function within ecosystem boundaries to sustain biological productivity, maintain environmental quality, and promote plant and animal health". Integrated soil quality indices based on a combination of soil properties provide a better indication of soil quality than individual parameters. Karlen *et al.* (1994) developed a soil quality index (SQI) based on four soil functions, namely the ability of the soil to: (1) accommodate water entry, (2) retain and supply water to plants, (3) resist degradation and (4) support plant growth. Each soil function was explained by a set of indicators. A similar framework was used by Glover *et al.* (2000) and Diack and Stott (2001). Harris *et al.* (1996) developed a SQI using three soil functions: the ability: (1) to resist erosion, (2) to provide plant nutrients and (3) to provide a favorable root environment. Hussain *et al.* (1999) modified this framework and demonstrated that adjusting threshold limits for local conditions can make the function ratings more or less sensitive to the management practices being evaluated. Kelting *et al.* (1999) used a soil quality model to measure the effects of management practices on forest soils by five growth determining attributes, namely the ability to: (1) promote root growth, (2) store, supply and cycle nutrients, (3) accept, hold and supply water, (4) promote gas exchange and (5) promote biological activity. Andrews *et al.* (2004) designed a "Soil Management Assessment Framework (SMAF)" for soil quality assessment, where indicators were selected based on the management goals, associated soil functions and other site-specific factors. They suggested six soil functions: nutrient cycling, water relations, physical stability and support, filtering and buffering, resistance and resilience, biodiversity and habitat; and three management goals: productivity, water recycling and environmental protection. All of the above function-based soil quality assessments were developed for use with temperate soils, whereas soil quality research on tropical soils, particularly in India, is limited (Sharma *et al.*, 2005).

1. Soil Quality Component or Indicators

A soil-quality indicator is a simple attribute of the soil can be measured to assess quality with respect to a given function. It is vital to select attributes that are appropriate for the goal, given the complex nature of the soil and the exceptionally large number of soil parameters that may be determined. As soil functions can not measured directly, a researcher uses various soil quality component or indicators to assess how well soil functions within the ecosystem boundaries or land use. Measuring soil quality is an exercise in which we identified soil properties that are responsive to management practices for achieving goal, affect or correlate with environmental outcomes, and are capable of being precisely measured within certain technical and economic constraints. Doran and Parkin (1996) clearly stated that there are three main categories of soil indicators *viz.*, chemical, physical and biological. These quality indicators determined through typical soil tests of physical, chemical and biological properties by employing standard procedure. For determination of soil quality index, four main steps followed: (i) define the goal, (ii) select a minimum data set (MDS) of indicator that best represent soil functions, (iii) score the MDS indicators based on their performance of soil function and (iv) integrate the indicator scores into comparative index of soil quality. For example, the average yield of any crop for each management practices/using ameliorant may define as the goal because the farmers like to get more productivity from each unit of fertilizer/ameliorant applied. Only those soil properties that showed significant changes may selected to have a representative minimum data set (MDS) (Andrews *et al.*, 2002a) through principal component analysis (PCA) as reported by Andrews *et al.* (2002a,b). Principal components having high eigen values and variables with high factor loading were assumed to be the variables that best represented attributes. The soil quality index states the fitness of soil for sustainable crop productivity.

2. Ideal Soil Quality Indicators

- ☆ Correlate well with ecosystem processes
- ☆ Integrate all the components of soil quality *viz.*, physical, chemical, and biological properties of soil and processes.
- ☆ Accessible to many users
- ☆ Sensitive to management and climate
- ☆ Components of existing databases
- ☆ Easily interpretable

3. Categories of Soil Quality Indicators

The selection of soil quality indicators depends upon the nature of the soil function under consideration. These soil attributes are classified in three broad groupings namely physical, chemical and biological indicators (Table 2.1). Many of the physical and chemical soil attributes are permanent in time (inherent parameters). Chemical attributes of soil health are correlated with the capacity to provide nutrients for plants and/or retaining nutrients for plants and or/retaining chemical elements or compounds harmful to the environment and plant growth. In contrast, biological and some physical attributes are dynamic and exceptionally sensitive to changes in soil conditions and in management practices (dynamic parameters). They appear to be very responsive to different agricultural soil conservation and management practices such as non-tillage, organic amendments, and crop rotation. The selection of soil indicator attributes should be based on: (i) land use, soil function, reliability of measurement, spatial and temporal variability, sensitivity to changes in soil management, comparability in monitoring systems, and skills required for the use and interpretation (Nortcliff, 2002). As USDA (2006) chosen various physical, chemical and biological soil indicators that are important for assessing soil quality as per defined goal (Table 2.1).

Table 2.1: Categorization of Soil Quality Indicators

Sl.No.	*Physical Indicators*	*Chemical Indicators*	*Biological Indicators*
1.	Soil moisture	Soil reaction	Soil organic matter
2.	Bulk density	Electrical conductivity	Earthworms
3.	Porosity	Soil salinity	Nematode population
4.	Infiltration rate	Sodicity (Sodium saturation)	Mycorrhizal association
5.	Aggregate stability	Nutrient recycling (available macro and micro nutrients)	Particulate organic matter
6.	Soil texture	Cation exchange capacity	Potentially mineralizable nitrogen
7.	Soil structure	Buffering capacity	Respiration
8.	Soil crusting	Carbonate content	Soil enzymes
9.	Soil compactness	Toxic elements	Total organic carbon
10.	Water holding capacity		Microbial diversity
11.	Water retention		Fatty acid profiling
12.	Drainage		Bio-ability of contaminates
13.	Hydraulic conductivity		Earthworms
			Nematode population

Soil Organic Matter

Organic matter, or more specifically soil organic carbon, transcends all three soil indicator categories and has the most widely recognized influence on soil quality. Organic matter is tied to all soil functions such as aggregate stability (physical), nutrient retention, availability (chemical) and nutrient cycling (biological) and is itself a strong indicator of soil quality. Soil organic carbon is the back bone of all the biochemical process occurs in soil which governed by the microbial activity. Microorganism used soil organic carbon as a carbon substrate for increasing their growth and development simultaneously increasing biochemical activities in soil. Some indicators are descriptive that can be used in the field as part of a health card whereas others must be measured using laboratory analyses.

4. Soil Quality Evaluation Method

A procedure in which using all physical, chemical and biological properties of soil as indicators was employed for establishment of the minimum data set for evaluation of soil quality as reported by Rezaei *et al.* (2006). The detailed procedures are given in the following section.

Indicator Selection

Selection of representative soil indicators is the key for soil quality evaluation. Ideally, those indicators which cover a wide range of soil functions and each may liable to affect soil quality directly considered as soil quality indicators (Wang and Gong, 1998). The Pearson correlation coefficient employed to determine the eligibility of dependent variables for inclusion of soil quality evaluation. Those soil properties which may not show a significant correlation (a Pearson correlation coefficient <0.20) with yields can be eliminate from the data list. Total organic carbon or soil organic carbon, soil pH, are the most frequent proposed soil quality indicators followed by available phosphorus, different indicators of water storage and bulk density are used more than 50 per cent of reviewed indicators sets. Texture, available potassium and total nitrogen are also used as chemical indicators > 40 per cent research schemes. In most of the research articles stated that at least one indicator from each soil properties *viz.*, physical, chemical and biological is included. However, soil biological indicators were missing from 40 per cent of reviewed minimum dataset. Soil physical indicators notably related to water storage were often proposed in the earlier assessment schemes. Among the soil chemical indicators, content of soil organic carbon, pH, electrical conductivity, total N, available N, P and K and cation exchange capacity were frequently used as chemical indictor than all other indicators. Similarly, microbial biomass carbon, soil respiration, N mineralization, dehydrogenase and density of earthworms were more frequently used as biological indicators among the other biological indicators. Recent publications suggested that indicators are applicable for many soil process (Bone *et al.*, 2010) *e.g.* earthworms serve as indicators for both water and nutrient cycling but most of the old publication lack a clear conceptual relationship between indicators, soil functions and ecosystem services (Lima *et al.*, 2013).

The adoption of novel soil quality indicators into minimum data set is of great interest if have distinct management goals for a particular situation. Recent development in soil science notably in soil biology gives promises for future soil quality assessment scheme. Because, soil organisms (fauna and flora) plays a central role in soil functioning. Therefore, adding biological and biochemical indicators can greatly improved soil quality assessment (Barrios, 2007). In addition, the assessment of biological indicators of soil quality is needed to link abiotic soil properties to soil functions in respect of biochemical and biophysical transformation carried out in above and below vegetation performance (Lehman *et al.*, 2015). Though, biological indicators of soil are still missing in soil quality assessments and mostly to black-box measurements like microbial biomass and soil respiration. Despite apparent potential of biological indicators, more specific indicators like nematodes and micro arthropods or soil biota have rarely been suggested (Stone *et al.*, 2016b), likely owing to require specific knowledge and skills. This condition is unlucky because soil biotas are considered the most sensitive indicators of soil quality owing to high effective to changes in environmental conditions. Also, there is an urgent need to include microbial diversity and structure of microbial community, soil-borne diseases and molecular methods of DNA and RNA as soil biological indicators. Molecular methods focusing on DNA and RNA hold great potential to perform faster, cheaper and more informative measurement of soil biota and soil processes as compared to conventional methods (Bouchez *et al.*, 2016).

Principal Component Analysis

Because of its MDS selection ability (Doran and Parkin, 1994), principal component analysis is used as a data-reduction tool to select the most appropriate indicators, through which the number of independent variables may be reduced and problems related to multi-collinearity could be eliminated. According to the MDS selection procedure as prescribed by Rezaei *et al.* (2006), only the principal component (PC) with eigen value > 1 may considered for identifying the MDS. Within each PC, indicators receiving weighted loading values within 10 per cent of the highest weighted loading can be select for MDS. When more than one variable was retained within a PC, the correlations among those indicators were examined to determine whether any of those variables can be considered redundant for the MDS. It can assume that highly weighted variables are highly correlated, and a linear correlation (r) greater than 0.70 may considered highly correlated. During the analysis, a VARIMAX rotation can be performed to enhance interpretability of the uncorrelated components (Flury and Riedwyl, 1988).

Weight Assignment

A soil quality index was computed by using the approach given by Karlen and Stott (1994), in view of three soil functions: (i) water infiltration, storage and supply, (ii) nutrient storage, supply and cycling, and (iii) sustained biological activity. All three soil functions were assumed to be equally important in this assessment and were assigned weight of 0.33. The 29 total soil quality indicators (Table 2.2) and the 8 indicators under MDS (Table 2.3) were included under three soil functions. In Table 2.2–2.4 indicators have been assigned to different levels. Under each level, numerical weights were assigned to soil quality indicators based on their assumed importance for the particular soil function under consideration. On the other way, the weight for each indicator are assigned by its communality after determined the final indicators for MDS, calculated by mathematical statistics of standardized factor analysis (Shukla *et al.*, 2006). The weight of each indicator may equal to the ratio of its communality with the sum of communalities of all indicators in the MDS.

Indicator Scoring

To avoid the effects of dimension and magnitude of different indicator units, a standard scoring function (SSF) (Andrews *et al.*, 2002a, b) may used to calculate soil indicator scores and normalize indicator measurements to a value between 0 and 1.0. Three types of SSF equations (such as S, reverse S, and parabola equations)

Table 2.2: Numerical Weight of 29 Indicators Set and Soil Functions Employed for Soil Quality Index Assessment (Lima *et al.*, 2013)

Soil Functions	*Weight*	*Indicator Level 1*	*Weight*	*Indicator Level 2*	*Weight*	*Indicator Level 3*	*Weight*
Water infiltration, storage and supply	0.33	Available water	0.20				
		Macroporosity	0.20				
		Earthworms	0.20				
		Correlated indicators 1	0.20	Soil organic matter	0.33		
		Correlated indicator 2		Bulk density	0.33		
				Microporosity	0.33		
			0.20	Water stable aggregates	0.50		
				Mean weight diameter	0.50		
Nutrient storage, supply and cycling	0.33	Available water	0.143				
		Macroporosity	0.143				
		Earthworms	0.143				
		Acidity/Al toxicity	0.143	pH	0.125		
		Correlated indicators	0.143	Potential acidity	0.25		
		Enzyme activity	0.143	Exchangeable Al	0.25		
		Nutrients	0.143	Aluminum Saturation	0.25		
				Soil organic matter	0.33		
				Cation exchange capacity	0.33		
				Microporosity	0.33		
				dippa-Glucosidase	0.33		
				Acid phosphatase	0.33		
				Alkaline phosphatase	0.33		
				Phosphorus	0.143		
				Potassium	0.143		

Soil Functions	*Weight*	*Indicator Level 1*	*Weight*	*Indicator Level 2*	*Weight*	*Indicator Level 3*	*Weight*
				Manganese	0.143		
				Mineralizable N	0.143		
				Copper	0.143		
				Zinc	0.143		
				Correlated indicators	0.143	Calcium	0.25
						Magnesium	0.25
						Iron	0.25
						Magnesium	0.25
Sustain biological activity	0.33	Microbial biomass	0.20	Soil respiration	1.00		
		Soil organic matter	0.20				
		Earthworms	0.20				
		Microbial activity	0.20				
		Microbial activity	0.20				

may be generally selected to standardize soil quality indicators (Cao and Zhou, 2008). The three SSF equations can be described as:

$$(1)\quad Type\ S: f(x) = \begin{cases} 0.1 & x - L \\ 0.1 + \dfrac{0.9 \times (X - L)}{(U - L)} & L \le x \le U \\ 0.1 & x > U \end{cases}$$

$$(2)\quad Types\ reverse\ S: f(x) = \begin{cases} 1.0 & x - L \\ 1.0 + \dfrac{0.9 \times (X - L)}{(U - L)} & L \le x \le U \\ 1.0 & x > U \end{cases}$$

$$(3)\quad \text{Type parabola}: f(x) \begin{cases} 0.1 & x < L_1 . x > U_2 \\ 0.1 + \dfrac{0.9 \times (x - L_1)}{(U_1 - L_1)} & L_1 \le x < U_1 \\ 1.0 - \dfrac{0.9 \times (x - L_2)}{(U_2 - L_2)} & L_2 < x < U_2 \\ 1.0 & U_1 \le x < L_2 \end{cases}$$

In these three equations, x is the monitoring value of the indicator; f(x) is the score of the indicator ranged between 0.1 and 1.0; and L and U are the lower and the upper threshold values of the indicator, respectively.

Table 2.3: Numerical Weight Associated to the Minimum Data Set and Soil Functions Used for Soil Quality Assessment

Soil Function	*Weight*	*Indicator Level 1*	*Weight*	*Indicator Level 2*	*Weight*
Water infiltration, storage and supply	0.33	Available water	0.25		
		Mean diameter	0.25	Soil organic matter	0.50
		Earthworms	0.25	Bulk density	0.50
		Correlated indicators	0.25		
Nutrient storage, Supply and cycling	0.33	Available water	0.25		
		Earthworms	0.25	Manganese	0.33
		Soil organic matter	0.25	Copper	0.33
		Micro-nutrients	0.25	Zinc	0.33
Sustain biological activity	0.33	Soil organic matter	0.50		
		Earthworms	0.50		

By using a scoring curve equation, three types of standardized nonlinear scoring functions typically can be generated: (1) "More is better" for positive slopes (*e.g.*, soil organic matter of soil organic carbon), (2) "Optimum" where a positive curve turns into a negative curve at a threshold value (*e.g.*, pH, EC) and (3) "Less is better" for negative slopes (*e.g.*, bulk density, toxic elements) and used for soil quality assessment as suggested by Karlen *et al.* (1994) and Glover *et al.* (2000). The shape of the curves generated by the scoring curve equation is determined by critical values. Critical values in which include threshold and baseline values (Table 2.5). Threshold values are values of soil properties where scoring function equals 1 (the best or upper threshold value) when the measured soil property is at an optimum level or equals

0 (poorest or lower threshold value) when the soil property is at an unacceptable level. Baseline values are soil property values where the scoring function equals to 0.5 and equals the midpoints between threshold soil property values (Glover *et al.*, 2000). Thus, numerical weights for each soil quality indicator are multiplied by indicator scores calculated through the use of standardized scoring functions that normalize indicator measurements to a value between 0 and 1.0. Scoring curves are generated from the following equation (Wymore, 1993):

$$\frac{1}{[1 + ((B - L)/(x - L))^{2S(B+x-2L)}]}$$

where B is the baseline value of the soil indicator (where the score equals 0.5), L is the lower threshold, S is the slope of the tangent to the curve at the baseline, x is the soil indicator value. Threshold and baseline values are generally based on literature, specific data reference, expert opinion or measured values observed under near-ideal soil conditions for the specific site and crop (Kelting *et al.*, 1999). For biological and physical indicators, threshold and baseline values may chosen according to the values from the data analyzed, because it is not possible to find a site that could characterize the natural soil condition in the region. For chemical indicators the threshold and baseline values were based on the literature (see Manual, 2004). This procedure is carried out for each group of indicators in each soil class studied (Table 2.4). In short, the soil quality index is rated (between 0 and 1.0) when all indicators for a particular function have been scored by summing the products of the numerical weights and the normalized soil properties scores.

Developing the Soil Quality Index

After all indicators in the MDS may scored and weighted, soil quality index (SQI) was calculated using the method described by Doran and Parkin (1994).

$$SQI = \sum_{i=1}^{n} Wi \times Si$$

where SQI is the soil quality index ranged between 0 and 1.0, Wi is the assigned weight of each indicator, Qi is the indicator score, and n is the number of indicators in the final MDS.

Various soil indicators used for agro-ecology or agriculture by various researchers are mentioned in Table 2.6. The values of various parameters were standardized between 0 and 1 using standardized scoring functions, but weighted scores were subjective and authors did not use any mathematical approach. As these authors stated single objective to show the methodology for computing a soil quality index, rather than provide a definitive tool for environmental evaluation. Karlen *et al.* (1994b) used SQI index to evaluate tillage effects on soil quality and exhibited that the absence of tillage improved soil quality. Karlen and Stott (1994) and Karlen *et al.* (1994a) used similar quality functions and method for developing soil quality index. These authors showed that the integrated production system lead the highest soil quality value. With a similar model to Karlen's and the function resistance to

Table 2.5: Scoring Function Values for Evaluating Soil Quality in Soil Textural Classes (per cent clay content). LT-Lower threshold; UT-Upper threshold; LB-Baseline; UB, Upper baseline; O-Optimum (Lima *et al.*, 2013)

Scoring Curve	*Indicator*	*Class 1 (<20 per cent clay)*					*Class 2 (20-40 per cent clay)*			*Class 3 (40-60 per cent clay)*		
		LT	*UT*	*LB*	*UB*	*O*	*LT*	*UT*	*LB*	*LT*	*UT*	*LB*
Physical properties												
More is better	Water stable aggregates	0	88	44			0	95	47.5	0	98	49
More is better	Mean weight diameter	0	3	1.5			0	5.0	2.50	0	5.60	2.80
More is better	Available water	0	0.14	0.07			0	0.09	0.05	0	0.08	0.04
More is better	Microporosity	0	0.36	0.18			0	0.48	0.24	0	0.61	0.30
More is better	Macroporosity	0	0.12	0.06			0	0.10	0.05	0	0.13	0.06
More is better	Bulk density	1.50	2.00	1.75			1.30	2.0	1.65	0.75	1.75	1.25
Biological properties												
More is better	Alkaline phosphatase	0	17	8.5			0	42	21	0	144	72
More is better	dippa-Glucosidase	0	57	28.5			0	122	61	0	161	80
More is better	Acid phosphate	0	206	103			0	350	175	0	506	253
More is better	Soil respiration	0	0.15	0.08			0	0.20	0.10	0	0.23	0.12
More is better	Microbial biomass	0	292	146			0	654	325	0	853	425
More is better	Earthworms	0	360	180			0	660	330	0	303	151
More is better	Microbial quotient	0	3.60	1.80			0	3.14	1.57	0	2.91	1.45
More is better	Potentially mineralizable N	0	47	23.5			0	64	32	0	72	36
Chemical properties (Functional values similar for all textural classes)												
More is better	Organic matter	0	5	2.5								
More is better	Total N	0	0.40	0.20								
More is better	Cation exchange capacity	0	15	7.50								
More is better	Ca	0	4	2								
More is better	Mg	0	1	0.5								

Scoring Curve	*Indicator*	*Class 1 (<20 per cent clay)*					*Class 2 (20-40 per cent clay)*			*Class 3 (40-60 per cent clay)*		
		LT	*UT*	*LB*	*UB*	*0*	*LT*	*UT*	*LB*	*LT*	*UT*	*LB*
More is better	K	0	180	90								
More is better	P	0	12	6								
Optimum	Fe	0	63	17.5	49	35						
Optimum	Mn	0	18	5	14	10						
Optimum	Cu	0	2.70	0.75	2.1	1.50						
Optimum	Zn	0	3.60	1.0	2.80	2.0						
Optimum	pH	3	8	4.20	7.0	5.50						
Less is better	Al	0	2	1								
Less is better	Al salt	8	40	16								
Less is better	H+Al	0	9	4.50								

biochemical degradation, Masto *et al.* (2007) established a soil quality index (SQI) with the objective to evaluate and quantify the long term effect of different fertilizer and farm yard manure treatments in a rotation system with maize. This SQI exhibited that soils subjected to inorganic fertilization and organic amendment treatments had higher SQI values than that of soils which were not fertilized or amended. A novelty of this study is that SQI calculate carried out without weighting factors which indicating that the use of these factors did not affect the relative ranks of individual treatments, as was found previously by Andrews *et al.* (2002a). Andrews *et al.* (2002a) were the first to compare multiparametric methods designed to establish the quality of agricultural soils, focusing on indicator selection methods: from opinion of experts (Doran and Parkin, 1994) to different mathematical–statistical systems. This mathematical development was later applied by different authors (Sharma *et al.*, 2005). The objective of this study (Andrews *et al.*, 2002a) was to evaluate the relative efficacy of various soil quality indexing methods using assessment of complex vegetation production systems in northern California. This work may be considered the mathematical base which many authors have since followed to establish indices in agricultural ecosystems. Moreover, Andrews *et al.* (2002a) also focused on the methodology for standardization and transformation of indicators from linear to non-linear methods, indicating that the latter requires deep knowledge about each indicator's behaviour and function within the system. Between linear and non-linear, the non-linear transformation was more representative of system function (Andrews and Carroll, 2001). Andrews *et al.* (2002a) retreated that both the expert opinion methods and statistical methods (principal component analysis) to select a minimum set of indicators is valid to establish the soil quality in vegetable production systems. The disadvantage of statistical methods resides on the need for a wide set of data, but, conversely, once the indicators are selected, there may be no need to assess soil quality over time. However, system and selected indicators could be more difficult to interpret by statistical approaches than expert opinion systems. Indicators selected through expert opinion (soluble phosphorus, pH, electrical conductivity, sodium absorption ratio and soil organic matter) varied from those selected by principal component analysis (soluble phosphorus, pH, calcium, sodium and total nitrogen). For most of soil quality indices established by these authors, organic systems received a higher SQI value as compared to conventional or low input treatments. Similarly, balance fertilization of mineral fertilizers recorded greater SQI in different crops than unbalance fertilization in Indo-Gangetic plains of India as reported by Singh *et al.* (2015), Singh *et al.* (2017) and Singh *et al.* (2018) (Table 2.6).

Conclusion

As a concept, soil quality has been widely defined and, perhaps for this reason, is far from being an objective measure. Many studies have showed the way by which analyzing soil quality using different soil quality indicators, its scores and weight but only a few have used the obtained results to establish a soil quality index. However, the few indices that exist have not been widely used. There are no universally applicable methods or formula to measure soil quality. It is very vital to not confuse an index with the methods used to establish it, but it based on the

Table 2.6: Indicators Used by different Authors in Soil Quality Indexes in Agro-Ecosystems

Authors	*Objectives*	*Soil Quality Indicators Used*
Karlen *et al.* (1994a)	Soil Quality Index: Evaluation ofeffects of crop residue management on soil quality under corn culture	Aggregate stability, porosity, worms, microbial biomass, respiration, ergosterol, total C, total N, bulk density, available water, pH, electrical conductivity.
Wang andGong (1998)	Relative Soil Quality Index: Evaluation of changes in soil quality in natural and agriculture systems	Soil depth, texture, slope, organic matter, total and bio-available N, P and K, cation exchange capacity and pH.
Hussain *et al.* (1999)	Land Quality Index: Adaptation of indices to evaluate effect of three cultivation systems on soil quality	Aggregate stability, organic C crop residues, porosity, exchangeable K, pH.
Liebig *et al.* (2001)	Agro-ecosystem performance: effects of conventional and alternative agricultural systems	Seed yield, N content of seed, pH, organic C, nitrates.
Kang *et al.* (2005)	Sustainability Index: Comparison of long term effect of organic amendments in systems for cultivating maize and rice.	Organic C, total N, extractable K, extractable nitrates and ammonium content, microbial biomass C and N, mineralizable N, respiration, bacterial counts, mycchorhizal infection, dehydrogenase activity.
Sharma *et al.* (2005)	Soil Quality Index: Selection ofadequate managements in drylands comparing between conventional and minimal cultivation.	Available N, K and S, microbial biomass C and saturated hydraulic conductivity.
Lee *et al.* (2006)	Soil Quality Index: Effects of swine manure compost application on soil quality under different vegetable and rice systems	Bulk density, aggregates, organic C, pH, available K and P, extractible Cu and Zn, microbial biomass, C, mineralizable N.
Puglisi *et al.* (2005)	Alteration index: Effects on the quality of agricultural soils contaminated with industrial and municipal wastes, organic fertilization or irrigation with poor quality water under different crops: Ficus carica, maize, tomato, *etc.*	Arylsulphatase enzymatic activities, β-glucosiadase, phosphatase, urease, ivertase, dehydrogenase and phenoloxidase.
Puglisi *et al.* (2006)	Soil Alteration Index: Effects on the quality of agricultural soilscontaminated with industrial andmunicipal wastes, organic fertilization or irrigation with poor quality water under different crops: F. carica, maize, tomato, *etc.*	PLFAs (Phospholipid fatty acid)
Mohantyet al. (2007)	Soil Quality Index: Effects ofcultivation practice (conventional and without ploughing) in rice–wheat systems, and maintaining vegetal residues on soil quality.	Bulk density, aggregate stability, resistance to penetration, organic matter

Authors	*Objectives*	*Soil Quality Indicators Used*
Masto *et al.* (2007)	Soil Quality Index: Evaluation ofagricultural soils fertilized withinorganic and/or farm yard manure.	Bulk density, water retention, pH, electrical conductivity, bio-available nutrients, organic matter, microbial biomass and crop yield.
Singh *et al.* (2015)	Soil quality index: Impact of balanced fertilization on nutrient acquisition, fibre yield of jute and soil quality in New Gangetic alluvial soils of India	Soil organic carbon, urease activity, *Azotobacter* counts, available P, microbial biomass carbon.
Singh *et al.* (2017)	Soil quality index: Identification of Minimum Data Set Under Balanced Fertilization for Sustainable Rice Production and Maintaining Soil Quality in Alluvial Soils of Eastern India.	Alkaline phosphatase, SOC, available K, phosphate solubilizing microorganism, available N
Singh *et al.* (2018)	Soil quality index: Effect of balanced fertilizers on soil quality and lentil yield in Gangetic alluvial soils of India	SOC, acid phosphatase, total culturable fungi, available, phosphorus, BMC

works of Karlen *et al.* (1994a, b) and Andrews *et al.* (2002a). Although the methods exist, indices are never used on a larger scale, nor even in similar climatological, agronomic *etc.* conditions. Because, most handicaps concerning their application are related with the poor standardization of the methods and with problems related with the spatial scale where they can be applied. In this sense, a clear definition of the conditions in which an index has been obtained must be kept in mind so that it can be applied at least at regional scale by the group concerned. In this case identifying and using some important site-specific factors for the study area, such as climate parameters and vegetation type and density are essential to decrease the poor of standardization and solve problems related to spatial scale. Moreover, a soil quality index should be defined use-dependent to some extent, so that it will be applicable in a larger scale. Thus, it will help us to select a minimum set of indicators which can address the maximum capacity of soil for specify function.

References

Alexander, M. 1971. Agriculture's responsibility in establishing soil quality criteria. pp. 66-71. In: Environmental Improvement – Agriculture's Challenge in the Seventies. National Academy of Sciences, Washington, D.C.

Andrews, S.S., Carroll, C.R., 2001. Designing a decision tool for sustainable agroecosystem management: soil quality assessment of a poultry litter management case study. *Ecol. Appl.* 11, 1573–1585.

Andrews, S.S., Karlen, D.L., Cambardella, C.A., 2004. The soil management assessment framework: a quantitative soil quality evaluation method. *Soil Sci. Soc. Am. J.* 68, 1945–1962.

Andrews, S.S., Karlen, D.L., Mitchell, J.P., 2002a. A comparison of soil quality indexing methods for vegetable production systems in Northern California. *Agric. Ecosyst. Environ.* 90, 25–45.

Andrews, S.S., Mitchell, J.P., Mancinelli, R., Karlen, D.L., Hartz, T.K., Horwarth, W.R., Pettygrove, G.S., Scow, K.M., Munk, D.S., 2002b. On-farm assessment of soil quality in California's central valley. *Agron. J.* 94, 12–23.

Barrios, E., Trejo, M.T., 2003. Implications of local soil knowledge for integrated soil management in Latin America. *Geoderma* 111, 217–231.

Bhatt R, Kukal SS, Arora S, Busari MA, Yadav M. 2016. Sustainability issues on rice-wheat cropping system. *Intern Soil Water Cons Res* 4(1): 68-83. http://dx.doi.org/10.1016/j.iswcr.2015.12. 001.

Bhattacharyya R, Ghosh B, Mishra P, Mandal B, Rao C, Sarkar D. 2015. Soil degradation in India: challenges and potential solutions. *Sustainability* 7(4): 3528-3570.

Biswas, B.C and Tewatia, R.K. 1991. Nutrient balance in Agro-Climatic regions of India-an overview. *Fertilizer News*, 36, 13-17.

Cao, Z.H., Zhou, J.M. (Eds.), 2008. Soil Quality of China. Science Press, Beijing, China (in Chinese).

Doran, J.W., Jones, A.J. (Eds.), 1996. Methods for Assessing Soil Quality. Soil Science Society of America Special Publication, vol. 49. Soil Science Society of America,Madison, WI, USA.

Doran, J.W., Parkin, T.B., 1994. Defining and assessing soil quality. In: Doran, J.W., Coleman, D.C., Bezdicek, D.F., Stewart, B.A. (Eds.), Defining Soil Quality for a Sustainable Environment. SSSA Special Pub., vol. 34. Soil Science Society of America, Madison, Wiaconsin, USA, pp. 3–21.

Flury, B., Riedwyl, H., 1988. Multivariate Statistics. A Practical Approach. Chapman and Hall, London, Great Britain.

Gathala MK, Ladha JK, Saharawat YS, Kumar V, Kumar V, Sharma PK. 2011. Effect of tillage and crop establishment methods on physical properties of a medium-textured soil under a seven-year rice-wheat rotation. *Soil Science Society of America Journal* 75:1851-1862.

Gill, J.S. 1992. Land use, conservation, management and development of land resources of Punjab, Chandigarh, Punjab. Report of Department of Soil Conservation and Engineering.

Glover, J.D., Reganold, J.P., Andrews, P.K., 2000. Systematic method for rating soil quality of conventional, organic, and integrated apple orchards in Washington State. *Agric. Ecosyst. Environ.* 80, 29–45.

Harris, R.F., Karlen, D.L., Mulla, D.J., 1996. A conceptual framework for assessment and management of soil quality and health. In: Doran, J.W., Jones, A.J. (Eds.), Methods for Assessing Soil Quality. SSSA Spec Publ., vol. 49. SSSA, Madison, Wisconsin, pp. 61–82.

Harris, R.F., Karlen, D.L., Mulla, D.J., 1996. A conceptual framework for assessment and management of soil quality and health. In: Doran, J.W., Jones, A.J. (Eds.), Methods for Assessing Soil Quality. SSSA Spec Publ., vol. 49. SSSA, Madison, Wisconsin, pp. 61–82.

Hillel, D. 1991. Out of the Earth: Civilization and the life of the soil. Univ. of California Press, Los Angeles.

Hussain, I., Olson, K.R., Wander, M.M., Karlen, D.L., 1999. Adaptation of soil quality indices and application to three tillage systems in southern Illinois. *Soil Tillage Res.* 50, 237–249.

Kang, G.S., Beri, V., Sidhu, B.S., Rupela, O.P., 2005. A new index to assess soil quality and sustainability of wheat-based cropping systems. *Biol. Fertil. Soils* 41, 389–398.

Karlen,D.L., Stott, D.E.,1994.Aframework for evaluating physical and chemical indicators of soil quality. In: Doran, J.W., Coleman, D.C., Bezdicek, D.F., Stewart, B.A. (Eds.), Defining Soil Quality for a Sustainable Environment. SSSA Special Publ., vol. 34. Soil Science Society of America, Madison,Wisconsin, pp. 53–72.

Karlen, D.L., Mausbach, M.J., Doran, J.W., Cline, R.G., Harris, R.F., Schuman, G.E., 1997. Soil quality: a concept, definition, and framework for evaluation. *Soil Sci. Soc. Am. J.* 61, 4–10.

Karlen, D.L., Mausbach, M.J., Doran, J.W., Cline, R.G., Harris, R.F., Schumman, G.E.,1997. Soil quality: a concept, definition, and framework for evaluation. *Soil Sci.Soc. Am. J.* 61, 4–10.

Karlen, D.L., S.S. Andrews, and J.W. Doran. 2001. Soil quality: Current concepts and applications. *Adv. Agron.* 74:1-40.

Karlen, D.L., S.S. Andrews, B.J. Wienhold, and J.W. Doran. 2003. Soil quality: Humankind's foundation for survival. *J. Soil Water Conserv.* 58:171-179.

Karlen, D.L., Wollenhaupt, N.C., Erbach, D.C., Berry, E.C., Swan, J.B., Eash, N.S., Jordahl, J.L., 1994. Long-term tillage effects on soil quality. *Soil Till. Res.* 32, 317–327.

Karlen, D.L.,Wollenhaupt, N.C., Erbach, D.C., Berry, E.C., Swan, J.B., Eash, N.S., Jordhal, J.L., 1994a. Crop residue effects on soil quality following 10-years of no-till corn. *Soil Tillage Res.* 31, 149–167.

Karlen, D.L.,Wollenhaupt, N.C., Erbach, D.C., Berry, E.C., Swan, J.B., Eash, N.S., Jordhal, J.L., 1994b. Long-term tillage effects on soil quality. *Soil Tillage Res.* 32, 313–327.

Katyal, J.C. 2003. Soil Fertility Management-a key to prevent desertification. *Indian Society of Soil Science*, 51:378-387.

Kelting, D.L., Burguer, J.A., Patterson, S.C., Aust, M.A., Miwa, M., Trettin, C.C., 1999. Soil quality assessment in domesticated forests— a southern pine example. *For. Ecol. Manag.* 122, 167–185.

Kukal, S.S and Aggarwal, G.C. 2003a. Puddling depth and intensity effects in rice-wheat cropping system on a sandy loam soil. I. Development of sub surface compaction. *Soil and Tillage Research*, 72: 1-8.

Lee, C.H., Wu, M.Y., Asio, V.B., Zueng-Sang, C., 2006. Using a soil quality index to assess the effects of applying swine manure compost on soil quality under a crop rotation system in Taiwan. *Soil Sci.* 171, 210–222.

Letey, J., R.E. Sojka, D.R. Upchurch, D.K. Cassel, K.R. Olson, W.A. Payne, S.E. Peterie, G.H. Price, R.J. Reginato, H.D. Scott, P.J. Smethurst, and G.B. Triplett. 2003. Deficiencies in the soil quality concept and its application. *J. Soil Water Cons.* 58:180-187.

Liebig, M.A., Varvel, G., Doran, J., 2001. A simple performance-based index for assessing multiple agro-ecosystem functions. *Agron. J.* 93, 313–318.

Lima, A.C.R. Brussaard, L., Totola, M.R. Hoogmoed, W.B., de Goede, R.G.M. 2013. A functional evaluation of three indicator sets for assessing soil quality. Applied Soil Ecology. *Applied Soil Ecology* 64: 194–200.

Manna, M. C., A. Swarup, R. H. Wanjari, H. N. Ravankar, B. Mishra, M. N. Saha, Y. V. Singh, D. K. Sahi, and P. A. Sarap. 2005. Long-term effect of fertilizer and manure application on soil organic carbon storage, soil quality and yield sustainability under sub-humid and semi-arid tropical India. *Field Crop Research* 93:264–80.

Manual, 2004. de adubac¸ ão e de calagem para os estados do Rio Grande do Sul e Santa Catarina, 10th ed. Sociedade Brasileira de Ciência do Solo, Porto Alegre. Mohanty, M., Painuli.

Masto, R.E., Chhonkar, P.K., Singh, D., Patra, A.K., 2007. Soil quality response to long-term nutrient and crop management on a semi-arid Inceptisol. *Agric. Ecosyst. Environ*. 18, 130–142.

Mohanty, M., Painuli, D.K., Misra, A.K., Ghosh, P.K., 2007. Soil quality effects of tillage under rice–wheat cropping on a Vertisol in India. *Soil Tillage Res*. 92, 243–250.

National Research Council (NRC). 1989. Alternative agriculture. Natl. Acad. Press, Washington, D.C. 448 pp.

Nortcliff, S., 2002. Standardization of soil quality attributes. Agric. Ecosyst. Environ. 88, 161–168. Roming, D.E., Garlynd, M.J., Harris, R.F., McSweeney, K., 1995. How farmers assess soil health and quality. J. *Soil Water Conserv*. 50, 229–236.

Oldeman, L.R. 1994. The global extent of soil degradation. p. 99- 118 In: D.J. Greenland and I. Szabolcs (eds.) Soil Resilience and Sustainable Land Use. CAB International, Wallingford, UK.

Oussible M, Crookstan RK, Larson WE. 1992. Subsurface compaction reduced root and shoot growth and grain yield of wheat. *Agron J*. 84: 34–38.

Pesek, J. 1994. Historical perspective. pp. 1-19. *In* J. L. Hatfield and D. L. Karlen (eds.) Sustainable agriculture systems. CRC Press, Boca Raton, FL.

Puglisi, E., Del Re, A.A.M., Rao, M.A., Gianfreda, L., 2006. Development and validation of numerical indices integrating enzyme activities of soils. *Soil Biol. Biochem*. 38,1673–1681.

Rezaei, S.A., Gilke, R.J., Andrews, S.S., 2006. A minimum data set for assessing soil quality in rangelands. Geoderma 136, 229–234.

Sharma, K.L., Mandal, U.K., Srnivas, K., Vittal, K.P.R., Mandal, B., Grace, J.K., Ramesh, V., 2005. Long-term soil management effects on crop yields and soil quality in a dryland Alfisol. *Soil Tillage Res*. 83, 246–259.

Shukla, M.K., Lal, R., Ebinger, M., 2006. Determining soil quality indicators by factor analysis. Soil and Tillage Research 87, 194–204. USDA, 2006. Soil Quality Institute. Natural Resources Conservation Service. Available at http://soils.usda.gov/sqi/. Retrieved on 20 February 2006.

Singh, S. R., Kundu, D. K., Dey, P., Saha, A. R., Mahapatra, B.S. 2017. Identification of minimum data set under balanced fertilization for sustainable rice production and maintaining soil quality in alluvial soils of Eastern India. *Communication in Soil Science and Plant Analysis*. 48(18): 2170-2192. doi.org/10.1080/00103624.2017.1408815.

Singh, S. R., Kundu, D.K., Tripathi, M. K., Dey, P., Saha, A.R., Kumar, M., Singh, I and Mahapatra, B. S. 2015. Impact of balanced fertilization on nutrient acquisition, fibre yield of jute and soil quality in New Gangetic alluvial soils of India. *Applied Soil Ecology*, **92**: 24-36.

Singh, S.R., Kundu, D.K, Dey, P., Singh. P, Mahapatra. B. S. 2018. Effect of balanced fertilizers on soil quality and lentil yield in Gangetic alluvial soils of India. *The Journal of Agricultural Science* (Cambridge) 156: 225-240. https://doi.org/10.1017/S002185 9618000 254.

Sojka, R.E., Upchurch, R.R., 1999. Reservation regarding the soil quality concept. *Soil Sci. Soc. Am. J.* 63, 1039–1054.

Tiwari, K.N., Sharma, D.N. and Tripathi, S. K. 1989. Salt affected soils of Uttar Pradesh, their reclamation and management (pp, 1-34), Kanpur, India, C.S.A. University of Agriculture and Technology.

Wang, X., Gong, Z., 1998. Assessment and analysis of soil quality changes after eleven years of reclamation in subtropical China. *Geoderma* 81, 339–355.

Warkentin, B.P., and H.F. Fletcher. 1977. Soil quality for intensive agriculture. Pp. 594-598. Proceedings of the International Seminar on Soil Environment and Fertility Management in Intensive Agriculture. Society of Science of Soil and Manure, National Institute of Agricural Science, Tokyo.

Wymore, A.W., 1993. Model-Based Systems Engineering. An Introduction to the Mathematical Theory of Discrete Systems and to the Tricotyledon Theory of Systems Design. CRC, Boca Raton, FL.

Yadav DB, Yadav A, Punia SS, Chauhan BS. 2016. Management of herbicide-resistant *Phalaris minor* in wheat by sequential or tank-mix applications of pre- and post-emergence herbicides in north-western Indo-Gangetic Plains. *Crop Protection* 89: 239-247.

Chapter 3

Role of Protected Cultivation Technology for Doubling Farmers Income through Horticultural Crops

Awani Kumar Singh and Naved Sabir

Principal Scientist
Centre for Protected Cultivation Technology (CPCT)
ICAR-Indian Agricultural Research Institute (IARI), New Delhi – 110 012

Introduction

Agriculture in India often finds itself confronted to various constraints under changing dimensions that are currently manifest in rapid climate change, soil degradation, population pressure, land fragmentation, water scarcity, marketing and entrepreneurship *etc.* Horticulture is fast emerging as one of the most dominant faces of Indian agriculture. India is second largest producer of the vegetable crops after China estimated to 146 MT from 8.4 mh area and highest producer of the fruit crops. There is little possibility of addition to the cultivable area though area under horticulture has been increased during the last two decades largely at the cost of other crops. Of all the horticultural technologies, plasticulture-cum-protected cultivation, is emerging as one of the most promising fields particularly in terms of higher productivity, quality, increased income, entrepreneurship *etc.* though the area coverage of about 60,000 ha alone appears abysmal. But the progress of area coverage under protected technologies in terms of per cent coverage is highly satisfying. The other major reasons for higher adoption include – promotion of hi-tech horticulture by public financial institutions. Although, despite indubitable

potential of protected cultivation, its spread has not been linked concomitantly with dissemination of technology, quality inputs, post-harvest management and processing units besides overall market linkages. Net result is that despite the technology performing exceedingly well by upto date farmers, majority of them keep facing challenges to cope up with the demands of hi-end horticulture. This article deals with overall gamut of technology, its potential, suitable areas and crops, tackling the challenges of biotic and abiotic stresses, market linkages and overall empowerment of farmers in terms of knowledge, input availability and use-efficiency, plant protection, post-harvest loss management, and market linkages *etc.* It can be safely presumed, that, if adopted, in right earnest, plasticulture in horticulture has immense potential in realizing the sustainable solutions for higher productivity, better livelihood and nutritional securityto the farmers who adopt the technology, particularly as clusters of hi-tech farming cooperates.

Plasticulture in horticulture includes protected cultivation-cum-covered cultivation under greenhouses, polyhouses, nethouses, tunnels, mulches *etc.* all of which include intelligent and exhaustive use of plastics in various forms depending upon the requirements, suitability of regions, climates, crops and technologies infrastructure or paraphernalia *e.g.* UV stabilized polythene sheets, insect-proof nets, shadenets, mulches, drip irrigation pipes, filters, tanks, plastic-lined water harvesting, *etc.* all are discretely made up of plastics.

What Agriculture Aspires Protected Horticulture Delivers?

Policy initiatives for doubling the farmer income, *'Pradhan MantriSinchaiYojana'* invoking 'more crop per drop' and *'Pradhan MantriRozgarYojana'* are some of the encores of the Government assertions in the quest for overall development of the nation including agriculture. Incidentally, plasticulture or protected cultivation or covered cultivation can makea cognizable and tangible contribution in all of these policy initiatives of the Govt., both directly and indirectly, in horticultural sector.

It is interesting to note that plasticulture including protected cultivation is apparently the only sector of agriculture which is a solution for most of the its inherent constraints into an ideal system which could address them asmodules of systematic and sustainable farming including healthy planting material, quality seed production, conservation and optimization and input use efficiency of water and other resources, precision farming, GAP, IPM, INM, Grafting, soilless cultivation besides minimized impaction of climate change aberrations *etc.* which all find convergence to strengthen farming into a hub of profitable horticulture. Apart from this, the hi-tech agriculture encourages better yields from fragmented land, entrepreneurship potential, GAP produce, export potential *etc.* It is by virtue of these strengths that protected cultivation technology can play a lead role in overall gamut of horticulture. Plasticulture is, thus, the future of horticulture. The potential of addressing soil erosion, recycling abilities, as well as amelioration of soil and water pollution are additional boons to overall ecosystem besides the economy of agriculture. Also, protected cultivation is most amenable to hi-tech supply chain networks in horticulture which ensures higher income to growers. Over and above of these advantages, plasticulture is very amenable to women workforce.

Basket of technologies covered under plasticulture are being discussed below with relevant details:

Scope and Advantages of Polyhouse Technology

Two types of polyhouses are designed in Indian condition, one is called naturally ventilated polyhouse and the other is called force ventilated (fan-pad) polyhouse. Naturally ventilated polyhouses are used mainly in temperate and moderate climate conditions while force ventilated (fan-pad) polyhousesare suitable for tropical, sub-tropical and arid zone farming systems. Polyhouse is main structure of protected -cum-covered cultivation groups. It is made by GI pipe and covered by ultra-violet (UV) stabilized plastic sheet with a service life of 3-4 years. These sheets are generally available in 7 and 9 meter widths with 200-micron thickness. Polyhouse technology is the most intensive form of commercial cultivation. Presently, various state Govts extend financial support in the form of subsidies to the growers limited to a ceiling of 1 acre of structures costing

approximately Rs. 38-40 lakh of which about 50 per cent of the cost is borne by Central Govt. with sizable and additional subsidies from State Govts which include the cost of structure, drip irrigation and fertigation system and seed/planting material *etc.* Polyhouse structures provide favorable growing conditions to crop plants and protect them from extreme fluctuations in weather, especially temperature, rainfall and wind and also minimizing evaporation lossesbesides protection from birds, animals and insect pests and diseases to quite an extent. The polyhouse structures provide 4-5 time higher yields with export quality produce; high water and fertilizer use efficiency and round the year production facility in any part of the country. Drip irrigation is mandatory for polyhouse crop cultivation along with fertigation technique because drip technology enhances water and crop productivity and minimized diseases, weeds, moisture fluctuations, irrigation frequency and judicial use of water and fertilizer. Thus, polyhouse and drip-irrigation technologies are components of horticulture based-precision farming system for better livelihood and nutritional security in the farming society.

Suitable Horticultural Crops for Polyhouse Farming and their Seasonal Rotations for Profitable Farming

Horticulture-based farming systems are highly diverse numbering more than 300 species, depending upon the climate, culture, season, size of landholdings, knowledge, market *etc.* Being high value crops with higher nutritional contents, besides with a higher demand in urban and peri-urban areas, horticultural crops provide a paradise for the progressive grower on one end and marginal farmer. Major crops for these systems are indeterminate tomato, cherry tomato, cucumber, capsicum, bitter gourd, musk melon, summer squash, vegetable nursery *etc.* in vegetable groups. Among flower crops, rose, gerbera, lilium, tulip, anthurium, carnations, orchids, chrysanthemum and their nursery. Among, fruit crops group, strawberry, papaya and all fruit-nursery propagation or hardening are used for economically profitable horticulture based farming systems.

Crop Diversity and Rotation for Horticulture Based Polyhouse Farming and their Impacts

Vegetables

After years of field experiences and feedback from the interaction with farmers following crops appear to give higher yield and better quality that ensure more income for improving livelihood and nutritional security. Wherein, indeterminate tomato hybrids (*GS-600, ID-32 and ID-37, Rakshita, Himsona, Himsikhar, Snehlata, Naveen etc.*) on an average give quality yield of 80-100 t/acre/crop with net returns nearing Rs. 3.0-4.0 lakh/acre/yr which is about 2-3 times higher yield and income as compared to traditional open farming systems. In the case of coloured capsicum (known as solanaceous apple) hybrids like *Natasha, Swarna, Indra, Bombi, Orobelle, Bachata, Inspiration etc.* in green, red and yellow colours produce 40-50 t/acre/crop yielding 6-7 times higher yield and income as compared to open cultivation of capsicum. Because the coloured capsicum cannot be grown properly in open condition. This crop, therefore, demands to be produced under protected horticulture. Among vegetables, cucurbits appear most suitable for protected cultivation both in terms of season and number of crops for better utilization of structures. 2-3 crops per year of parthenocarpic cucumber (cv. Isatis, Kian, Hilton, Sun Star, Multistar, Fadia, Mini Angel *etc.*), summer squash (cv Pusa Alankar, Pusa Pasand, Australian Green, Seoul Green, Kora, Yellow Zucchini, Himanshu, *etc.*), bitter gourd (Pusa Rasdar) or musk melon (Pusa Sarda) and yellow musk melon can be taken in polyhouses during early and off-seasons. Average production of 120-150 t/acre/year can be obtained in cucumber. These vegetables are transplanted during July to Aug. and harvested from May to June in the plain conditions. However, in hill areas, transplanting is done in the month of Feb.to March and harvesting during Oct. These vegetables result in better livelihood and nutritional security as compared to traditional vegetable farming in the open.

Cropping-Sequence-cum-Rotation-cum-Cropping-Intensity of Vegetable Crops for Polyhouse Farming

Different Crop Cycles of Polyhouse Vegetables in a Year	*Returns/1000 sq.m. (Rs. in lakhs)*
Year round Veg./fruits/flower crops nursery	5-10
Veg. Nursery + Cucumber or Capsicum or Tomato or bitter gourd or Musk melon or long melon	3-4
Spinach or Coriander leaf + Tomato (or Capsicum or Cucumber)	2-3
Cucumber + Tomato (or Capsicum)	2-3
Spinach or Coriander leaf + Cucumber + Cucumber + Cucumber	3-4
Cucumber + Strawberry + Cucumber (or Bitter gourd/Muskmelon/Leafy vegetable)	4-5
Veg. nursery or Spinach or Coriander leaf + Cucumber + Lettuce + Musk Melon	3-4
Spinach or Coriander leaf + Bitter gourd + Lettuce + Musk melon	3-4
Spinach or Coriander leaf + Bitter gourd + Lettuce + Bitter gourd	3-4
Spinach or Coriander leaf + Musk melon + Lettuce + Musk Melon	4-5
Spinach or Coriander leaf + Kakri + Lettuce + Musk melon	3-4
Spinach or Coriander leaf + Summer squash + Lettuce + Musk Melon or Cucumber	3-4
Nursery + Spinach or Coriander Leaf + Chilli or Brinjal	2-3
Veg. Nursery + Lettuce or Spinach or Coriander leaf + Cucumber or Musk melon	2-3
Vegetable Nursery + leafy vegetable + French bean (or okra) + Cucumber (or Musk Melon)	2-3

Flowers

Polyhouse flower crops like rose, gerbera, tulip, Anthurium, carnations, orchids, lilium, and chrysanthemum and their nursery are mostly annual or perennial crops and hence seasonal crop rotations are not required. Most of these cut flowers required round the year as per market demand can be cut. Floriculture crops are highly remunerative as the quality, colourand shelf-life as well as freshness is very important and can give annual returns of 8-10 times more than the open field conditions. They are also more important from export perspective, so critically important in the country. Floriculture industry is led by produce from polyhouse growers mainly in Europe and developed nations. Thus, floriculture can ensure more income which in turn promotes better livelihood and nutritional security of farming families undertaking horticulture-based farming.

Fruits

Major fruit crops recommended underpolyhouse cultivation include papaya, strawberry, mango, guava, pomegranate and propagation materials of high value orchards or breeding material. These fruits crop can be cultivated for fresh and disease-freeseedlings or plantproduction. Papaya fruits and seeds arein high demand in the market. Best quality papaya fruits yield on average 45-50 kg/plant with net returns of about to Rs. 5-6 lakh/acre. However, strawberry fruit yield of15-20 tons/acre are harvested in the 10 plants per sq. meter. All fruit crops are more effective in minimizing malnourishment being nutrient rich. Strawberry runners, virus-free papaya seed, budded and grafted plantsand meadow orcharding fruits are moreeconomicallyand safe produced for livelihood and nutritional security. papaya can be cultivated in polyhouse as inter-cropping withleafy vegetable crops likespinach, coriander, lettuce, mustard or French been and or strawberry crops. Strawberry transplanted in the August-September and harvested up in April followed by papaya transplanted in twice, either in July to August or later in Jan-Feb month per year. Crop rotation can be used withstrawberry crop *i.e.* leafy vegetable or vegetables nursery or cucumber + Strawberry + cucumber (or bitter gourd or muskmelon or leafy vegetable) can be taken.

Crop Diversity and Rotations for Horticulture Farming under Net-Built Structures and their Impacts

Generally two types of net-built structures are used *viz.*, shade-net and insect-proof net houses often termed synonymously, but indeed, a shadenet is primarily a covering of UV stabilized coloured plastic net where main objective is to provide protection to the horticultural nurseries and plants from harsh sunlight and hot winds during summer seasons. On the other-hand, insect proof nethouses are built for the primary objective of screening out the entry of insects or other pests by the physical barrier of nets of various mesh sizes depending upon crop and season. Both the nethouses can be made by iron or GI pipe or bamboo or wooden or stone pillars *etc.* and are one of the most feasible forms of protected cultivation. Basic cost of fabrication of these structures may range from Rs. 8-10 lakhs/acre.

Shade Net Protected Structures

Shade nets generally use two colour nets (green and black) in horticulture-based farming systems. Both colour shade nets are available in 30, 50, 75 and 90 per cent capacity of shading protection, wherein this percentage means the percentage of reduction in sunlight. In vegetable crops 30-50 per cent capacity of either colour or 50-90 per cent shade nets are used for fruit-propagation in horticultural crops. Shade nets also cut down the sunlight, in turn minimizing evapotranspiration losses of moisture in general as well as reduced transpiration from plants. The total effect creates an altered relative humidity and temperature regime of microclimate favouring the plant growth. That is why these structures are especially useful for dryland and arid areas with water scarcity by increasing water use efficiency when used in combination with drip irrigation. Shading nets are the perforated plastic materials used to provide relief to theplants from the scorching sunlight, hotwinds, direct rainfall, hails, birds and animals as well as insects-pests. Nets of 25 per cent to 75 per cent shading capacity are beingused for raisingnursery, indoor plants, hardening of tissuecultured plants and growing of vegetables. During high light intensities, leafy vegetables and ornamental greens are recommended to be grown under shade nets. This structure enhances 25-35 per cent more yield as compared to traditional methods during peak summer season. Suitable crops identified for shade nethouses from economic viability include horticulture nurseries, leafy vegetables, summer tomato and capsicum. Because these systems are mainly used to grow off-season cultivation of crops, their produce fetches much higher prices in the market, hence providing good economic returns to growers, thereby improving livelihood and nutritional security to the farming families. Ornamental crops like Lilium, tulip, anthodium and orchids can also be grown economically depending upon suitability of different agro-climatic zones. Therefore, shade nets have come to occupy a prime importance in horticultural promotion in the country especially vide propagation of quality planting material.

I.P. Net Protected Structures

Insect-proof nylon net based cultivation is newer concept more suitable for rainfed areas with much fewer days of precipitation. These structures make use ofinsect proof screens of different intensities of perforations, ranging from 25 mesh to 60 mesh. IPP nets of 40 or higher mesh are effective means to control entry of most flying insects and save crop from diseases. As compared to polyhouses, these structures are almost equally capable of screening out the insects and birds thereby avoiding the use of pesticides. These structures permit early planting of tomatoes and cucumber, bitter gourd, musk melon, summer squash, brinjal and chilli *etc.* without the risk of vectors. Higher mesh size (*i.e.* above 50), however, reduces the air exchange of the structure. UV- stabilized white coloured nets are now available which have a longer life. This structure enhances 25-35 per cent more yield as compared to traditional methods during peak summer season. IP net structures can be used successfully for seed production in rainfed areas. Crops suitable for these structures include tomato, cherry tomato, cucumber, bitter gourd, musk melon, summer squash, papaya, meadow and high density orcharding of mango, guava and pomegranate and strawberry *etc.* besides very few flower crops. In irrigated areas, these structures may face problems during rainy season as it tends to flood the crop inside. Hence it is advisable to cover its roof by polythene sheets during rainy period. But this is possible only with small, half-moon type or gable type structures. Insect-proof covered cultivation technique give 2-3 times more yields and hence increase the income by about 30 per cent normally but even more at times where insect damage is more. Thus, farmers are able to save the crop hence being safer for farmers.

Plastic Mulching

Plastic mulching technology, also called surface covered cultivation, largely involves mulching with polythene sheets for addressing three major issues *viz.*, soil and water conservation, leaching of nutrients as well as reduction of weeds and to some extent insect pest damage. Drip irrigation is an integral part of mulching technology. Mulching involves covering the soil around the plant bases with an organic or inorganic material which makes condition more favourable for plant growth and development. Organic mulches, like leaves, straw, sawdust *etc.* add nutrients and humus to the soil as they decompose, improving its tilth and moisture holding capacity. Synthetic or plastic mulches have various beneficial effects on crop production. Plastic mulch accelerates plant growth by increasing soil temperature, conserving soil moisture, weed control, production of quality produce and reduction in leaching of nutrients. The plastic mulch is available in different colours. Each colour has its own significance. Transparent polyethylene mulch raises the soil temperature by soil solarization used mainly for managing soil-borne diseases and nematodes. This effect derives mostly from the suppression of latent heat loss through evaporation. Black polyethylene film also gives effective weed control by cutting down solar radiation by more than 90 per cent, resulting in etiolated growth and the eventual death of weeds under the film. The yellow plastic mulch attracts insectsso it can be used toattract and kill insects. The silver-reflective type of plastic mulch associated with higher reflectance causes insect disorientation and repels aphids. The different types of mulches are reflective plastic mulches, infra-red transmitting mulches and biodegradable plastic mulches. There are different types of laying methods for plastic mulch but large area usage has to involve the help of mechanised and tractor operated mulch layers. Similarly, several low-costhole cutting devices and transplanting devices are now pouring in market. Suitable horticultural crops that can take advantage of mulching include plantsgrowing vertically by nature and not the creeping types *e.g.* solanaceous crops and cole crops. Although, cucurbits like summer squash can also be grown on mulching but with proper staking systems only for taking better advantage of the strengths of this type of cultivation. In fruit crops also the mulching is very important especially in strawberry and gooseberry. Other fruit crops like papaya, grapes, pomegranate, guava, mango *etc.* under high density and meadow planting methods do use the mulching technology strategically. This technology is again useful for areas with water scarcity or rainfed ecosystems, and can conserve moisture to the extent of 50 per cent water saving and 30 per cent nutrient saving. Thus, we can conserve resources, economise input use and efficiency and harvest 25-30 per cent higher yields as also the quality, playing a significant role increase in income and overall upliftment of livelihood of farming families.

Hi-Tech Techniques for Healthy Planting Materials for Horticultural Crops

Plasticulture based healthy planting material is the backbone of overall gamut of horticultural production systems. Nursery raising requires hi-end greenhouses with boom irrigation, glazing material, seeders, trays and cell pack for propagation, container carrying trays, polytube, bags, cups and labels for proper identification, ground cover and packing of planting material for long distance transportation.

As it is highly important to grow disease and pest free healthy planting material, soilless production of nurseries under greenhouses is very important. Generally plastic nursery trays or pro-trays having different cell sizes are used for raising vegetable and fruit seedlings. Different kinds of plastic pro-trays are used in raising flower seedlings. These trays help in proper germination of seeds and cuttingsand provide independent area for each seed and cutting to germinate, reduce the mortality rate, maintain uniform and healthy growth of seedlings, are easy in handling and storing, reliable and economical in transportation. These plastic trays may be fixed in thermocol base trays having the same number and size of cavities before filling the media. Media used are mainly two types, *viz.*, soilless media (3 parts of cocopeat + 1 part each of vermiculite and perlite) which costs about Rs. 6000-7000 per 100 kg or indigenous media (2 parts cocopeat+1 part vermicompost/ leaf compost and the other combination may include 2 parts vermicompost/leaf compost and one part each of sand and ash. These media may be enriched with

Trichoderma). The cost of these media is only Rs. 500-1000 per 100 kg. This media, is not only cheaper but also has base nutrients available.

Overall, hi-tech nursery is a little capital and labour intensiveor automation based yet highly remunerative to the extent that it can be adopted in entrepreneurial model byunemployed educated youth.

Rainwater Harvesting of Polyhouse Roofs with Plastic Lined Tanks

Rainwater harvesting of Polyhouse Roofs for storing in plastic lined tanks at proper place for long duration so that even a single drop of water can be utilized preciously. This scheme is being subsidy and training from Govt. of India under scheme of '*Pradhan Mantri Sinchaee Yojana*'. In present scenario, concept of poly-tank cum micro-irrigation system can be employed in protected cultivation. Special polythene sheets of 300-400 gsm thickness, UV stabilized, cross-laminated in blue or black colour are utilized as standards. Life of water storage tanks so devised extends beyond 10 years easily. Additional advantage of these systems is their ability to withstand earthquakes. Besides compensating the water scarcity, rainwater harvesting from polyhouse roofs can remediate water quality in areas where soil salinity and pH are inextricable issues. Such system is also the best way to provide irrigation, under low water availability areas like lower to high hills with high water facing acute water shortage in spite of rainfall. Rainwater so harvested can be best used for life-saving irrigation strategies. The micro-irrigation system saves water (40-50 per cent), fertilizer (25-30 per cent), labour expenditure (50 per cent), soil-erosion cum fertility (90 per cent), minimizes plant mortality (80 per cent), enhances crop yield (25-75 per cent) and improves the quality of produce as compared to traditional pattern specially when employed under protected environment. Keeping this in view, storage of rainwater in low cost adequate capacity poly-tank and use of this water for irrigation in crop grown in protected environment (Polyhouse or Tunnels) through gravity based drip irrigation system. Cost of installation of this type of poly-tank make is Rs. 120-150/ m^3 including roof top, lateral pipes and storage tanks. If drip irrigation is integrated with these structures which in turn give additional advantage to the water-scarce production system. This can even lead to increase in area under cultivation for these areas. Water storage tanks can be additionally used for pisciculture or duck-fish or hen-fish farming in integration with horticultural production system as these cultures add nutrients to the water, minimizing nutrient requirements besides getting extra income from meat, fish, eggs *etc.* for livelihoodbetterment. Thus, this technology can play good role in livelihood cum nutritional security for difficult areas through horticulture based cropping system. This system has been amply demonstrated at KVK, Lohaghat, Champawat, Uttarakhand and CPCT, IARI, New Delhi.

Micro-irrigation-cum-Fertigation Technique

Micro-irrigation is classified mainly into two types, i) drip irrigation and ii) sprinkler irrigation. Drip irrigation is one of the best available tools made out of plastics in agriculture for the judicious use of water for growing various crops in large scale on sustainable basis. Drip irrigation is a low, labour intensive and highly efficient system of irrigation, which is also amenable to use in difficult situations and problematic soils, even with poor quality water. This technology is the practical example of *"More Crop PerDrop"* theme of agriculture in PMSY and has provision for 50-60 per cent subsidy for the adoption of technology. Irrigation water savings through drip irrigation can be affected by adopting a suitable drip irrigation system ranging from 36-79 per cent. In drip irrigation, water is supplied through a network of plastic pipes using dripper/emitters. Water is supplied at a regular interval and at a required time/ quantity. Productivity gains vary from 20 per cent to 50 per cent depending upon the crop. Drip irrigation works by applying water slowly, directly to the soil. Drip irrigation has high efficiency as there is no loss of water in form of evaporation or run- off and water is applied directly to the plant roots. This is one of the major tools of precision farming which effectively increase water and nutrient use efficiency thus contributing significantly in water scare areas, increasing production and overall livelihood upliftment through horticulture based crops. Similarly, sprinkler irrigation is one of the more effective tools of water conservation. It is used mainly in production of horticultural crops in the open type of cultivation mainly orchards. This is used mainly in carpet natured crops and high density crops and leafy vegetables like beans, peas, root based crops like radish, carrot *etc.* and ornamental landscapes and lawns including early transplanted stages of fruit crops. This irrigation technique has given a flip to the intercropping system in horticultural crops. Currently, about 69.5 mha area is under micro-irrigation, of which drip irrigation covers about 27 mha. Lately, other cropping systems have also started to adopt the drip irrigation to finally taking agriculture to overall aim of rainbow revolution.

Farm Value Addition for Horticultural Produce including Packaging

In vegetable crops, around 28-30 per cent losses take place during harvesting, grading, packaging, transportation and marketing. The existing post-harvest loss of vegetables can be considerably reduced by adopting improved packaging, handling and efficient system of transport. Plastic is used for packaging and storage of vegetable crops for extending storage life. The plastic bags, plastic sacks, plastic blow moulded containers are used for handling and storage. The cellulose film (cellophane), cellulose acetate, polyethylene (HDPE and LDPE), polystyrene, ethyl cellulose, polypropylene, PVC, edible plastic is used for packaging of vegetables. These technologies, minimize losses and add value, fetch higher price and also enhance the shelf-life of horticultural produce.

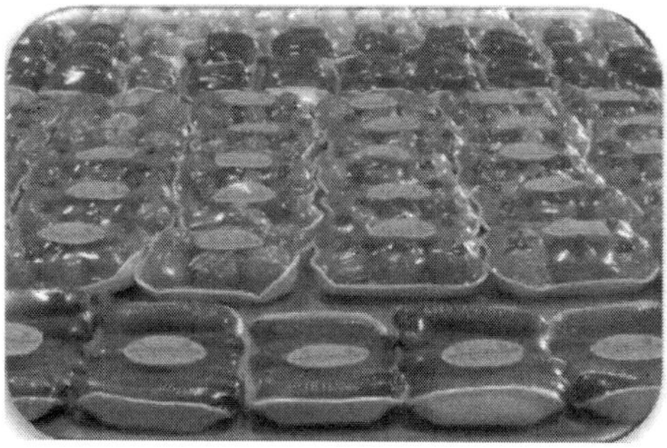

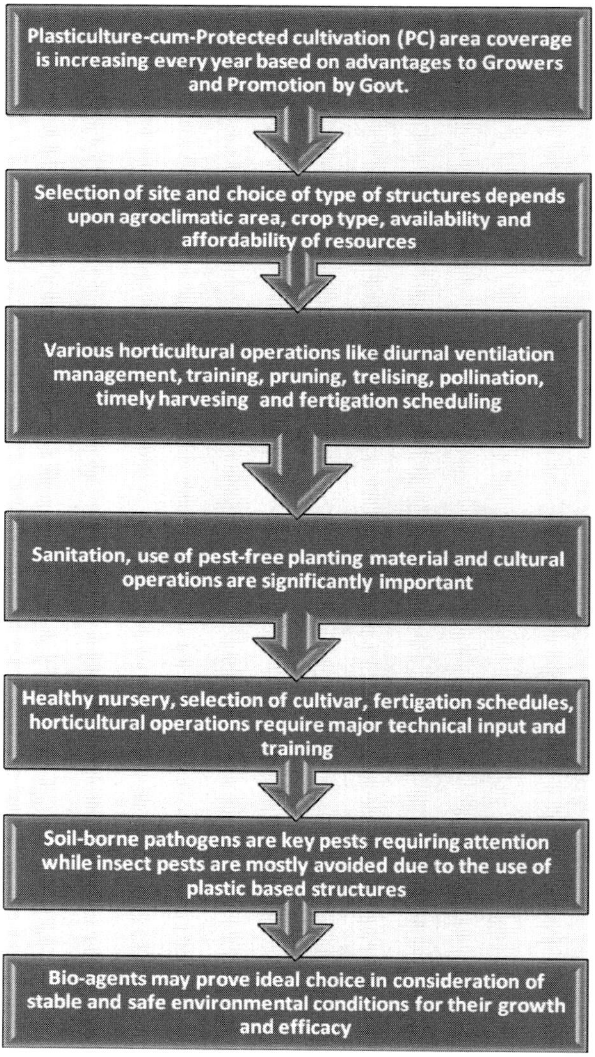

Hi-tech horticulture farming systems, thus, have immense uses in order to conserve the water, nutrient and energy resources besides increasing the production, generating additional income for the farmers, creating more employment opportunities in rural areas for the unemployed youth and women folk. In future, there is need to encourage environmentally degradable material, developing plastic based system for storage and handling of vegetable produce, besides adopting good agricultural practices and IPM. There is need of developing technology for production of eco-friendly plastic and eco-friendly disposal of waste plastic.

Chapter 4

Doubling Farmers Income through Farming System Approach

B. Gangwar

Former Director
ICAR-Indian Institute of Farming Systems Research,
Modipuram, Meerut – 250 110

Indian Agriculture has onus of providing national as well as house hold food and nutritional security to its teaming millions. However, the national sample survey organization points out that more than 40 per cent of farmers are willing to leave farming if better off farm employment is provided. This is mainly due to low rate of return and higher cost of inputs and human power. The strategies employed during green revolution era in seventies were mainly focused on increasing the production and productivity with emphasis on new high yield varieties, increased fertilizer use and irrigation. But, in recent years many second generation problems were experienced.

Through India has made a phenomenal increase of 5.5 times in food grain production from 51 MT in 1951 and a all time record of 284.8 MT of food grains production could be achieved during 2017-18 in India. On the other hand, the fact remains that the Indian farmer's income could not grow in the same pace as it could happen in others sectors of Indian economy. The National Sample Survey Organization data on consumption expenditure survey for the years 2011-12 reveals that more than one fifth of rural households with self employed in Agriculture as their principal occupation were having income less than the poverty line. The highest incidence of poverty has been reported in Jharkhand wherein 45.3 per cent population is living below poverty line as reported by Chand Ramesh 2017 (Figure 4.1) In India at present, an income of Rs. 45/- per day per head in rural areas and Rs.75/- per day per head in urban areas considered to be below poverty line. The majority number understood to be in rural areas directly depending on agriculture.

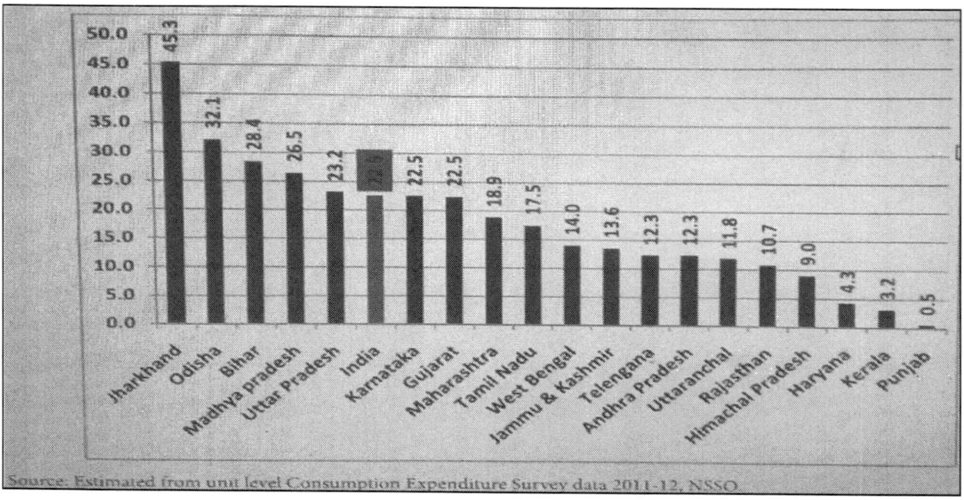

Figure 4.1: Farm House Hold Income below Poverty Line (2011-12). [*Source*: Ramesh Chand (2017)].

The Indian agriculture is dominated by small farm holders. In fact the smallholder farmers of developing countries are facing unprecedented challenges in the 21st century. With an estimated 9.2 billion people to feed by 2050 of whom 8 billion will be in developing countries including India and increasing scarcity of land and water, productivity gains will have to be the main source of growth in agriculture and the primary means to satisfy increasing demand for food and other agricultural products. With globalization and new supply chains, farmers will need to continuously innovate to respond to changing market demands and remain competitive. Moreover, "climate change has the potential to irreversibly damage the natural resource base on which agriculture depends. Climate adversity is increasing production risks in many farming systems and reducing the ability of farmers and rural communities to manage these risks on their own. Around the world, resource-poor farmers are trying to adapt to the effects of climate change, which affect them disproportionately: (i) dwindling crop yields; (ii) desertification and land degradation processes, exacerbated by changes in rainfall patterns; (iii) rising sea levels, affecting in particular the livelihoods of coastal communities; (iv) diminishing natural resource productivity; and (v) in some areas, irreversible loss of biodiversity. Planned farming activities has strong potential to reduce greenhouse gas (GHG) emissions by promoting clean and efficient energy, reducing deforestation and promoting sustainable farming system practices such as the rehabilitation of degraded lands, water conservation and management, and increased biomass production.

The farming system approach is considered to be the powerful tool for addressing above issues and ensuring doubling farmers income as being focused and targeted by the Indian Government in near future (2022-23). A judicious mix of two or more farm enterprises with advanced agronomic management tools may compliment the farm income together with help in recycling the farm residues. The selection of enterprises must be based on the principles of minimizing the competition and maximizing the complementarity between the enterprises. In general, farming system approach is based on the following objectives:

- ✰ Sustainable improvement of farm household systems involving rural communities.
- ✰ Farm production system improvement through enhanced input efficiency.
- ✰ Raising the family income.
- ✰ Satisfying the basic needs of farm families.
- ✰ Climate resilient/environment friendly.

Procedures for FSA

The five-step procedure for farming system approach is explained below

- ✰ **Classification:** Classification is concerned with the identification of homogenous group of farmers with similar natural and socio-economic characteristics. It forms the basis for the setting of priorities and for targeting of research, extension and development to particular farm types.
- ✰ **Diagnosis:** Diagnosis has to do with identifying the limiting factors, constraints and development opportunities of particular target farm types priority set as per resource availability.
- ✰ **Experimentation and recommendation:** Recommendations made from the the knowledge, but in field situations it involves experimentation, either at the farm level or at the research station or at both, as a pre requisite.
- ✰ **Implementation:** Implementation commitment is usually found in farming systems programmes directly through support to the extension agencies.
- ✰ **Evaluation:** Evaluation is an important component and will lead to reappraisal.

Coordinated Efforts

Indian Council of Agricultural Research (ICAR) through Indian Institute for Farming Systems Research (IIFSR) is implementing the Integrated Farming Systems Research from 2010. The IIFSR is also having the AICRP on Integrated Farming Systems with 75 centres. On-station (44) centres are working on the farming systems model development for small holders mainly for experiential learning, demonstration and modeling the resource flow. The on-farm (31) centres located in the districts are working directly with the farmers through module based interventions in farmer participatory mode. In each location 30 marginal holders and 6 small holders are covered for farming system model development. The coordinated research in farming system is essential mainly for looking the options for constraints from various regions, replicating the successful model of one region to other region and for development of resource centric models. Apart from the AICRP on IFS, a network project on organic farming is also operated by

IIFSR through 20 centres. Further, ICAR is also considering the farming systems approach of delivering technologies to the farmer door steps. All the time tested technologies generated by commodity based research institutions is being brought under the umbrella of farming systems and delivered to the farmers through its AICRP on IFS.

Pre-dominant Farming Systems

The quick survey conducted as a part of characterization of existing farming systems throughout the country indicates existence of 19 pre-dominant farming systems with majority as crop + livestock (85 per cent). Although crop + livestock system is dominating in the country, based on the per cent contribution to net income, the systems are classified as crop, horticulture, livestock, fisheries dominant systems where in dominant component contributes more than 50 per cent of the total net returns. Accordingly, it was found that crop dominant farming systems are existing in the states of Andhra Pradesh, Bihar, Chattisgarh, Goa, Haryana, Jammu and Kashmir, Jharkhand, Kerala, Karnataka, Madhya Pradesh, North-East, Maharashtra, Orissa, Punjab, Tamil Nadu, Uttar Pradesh and Uttaranchal while livestock dominant systems are present in Rajasthan and Parts of Gujarat. West Bengal, parts of Odisha and Assam states have the fisheries as a major source of income to the existing farming systems. The scope for promotion of horticulture (fruit) based systems exists in Jammu and Kashmir, Himachal Pradesh, Maharashtra, parts of Uttar Pradesh and in Sikkim while plantation dominant systems are available in Andaman and Nicobar Islands and Kerala. In selected states and locations, highly diversified systems also exists where in none of the component contributes for 50 per cent or more to the returns. The characterization survey data of existing farming systems in different states/zones are being interpreted and brought out in the form of National Atlas on Farming Systems. Though the various farming systems exists in the country, integration of input and output within the system is either completely lacking or at partial integration. Competition exists within and outside the farm for various byproducts generated. Cow dung is the best example as dung is required for improving the fertility of soil and meeting the household fuel. Hence, synergy needs to be made by appropriate allocation for making farming as a profitable option. Sustainable farming systems should aim for long term productivity, profitability, recycling of resources and employment generation.

Under on-farm research component of AICRP on Integrated Farming Systems, performance of existing farming systems in 732 marginal households in the country was taken up in 30 districts in 20 states and it clearly reveals that existence of 35 types of farming systems with components as high as 5. Out of the sample surveyed and sub group of crop+ livestock system, crop + dairy is the major system practiced by 48 per cent of marginal holders followed by crop + dairy + goat (11 per cent). Among the livestock category, dairy is practiced by 86 per cent marginal holders followed by goat (24 per cent) and poultry (21 per cent). The other components such as fish, fruits, apiary, sheep *etc.* are location specific. Further analysis of number of enterprises present in the farming system indicates, 52 per cent households are having two (example crop+dairy), 28 per cent farm households are having 3 (example crop+dairy+goat) and 11 per cent households are having 4 (example crop+dairy+goat+fish) (Figure 4.2). Around 7 per cent households are having the single enterprise (either crop or dairy alone).

The contribution of crop and livestock to gross income of marginal households in various NARP zones indicates that in majority of the place crop component contributes > 50 per cent while at few districts such as Samba (Jammu), Aurangabad (Maharashtra), Mehsana and Panchmahal (Gujarat), livestock component contributes either equally or more.

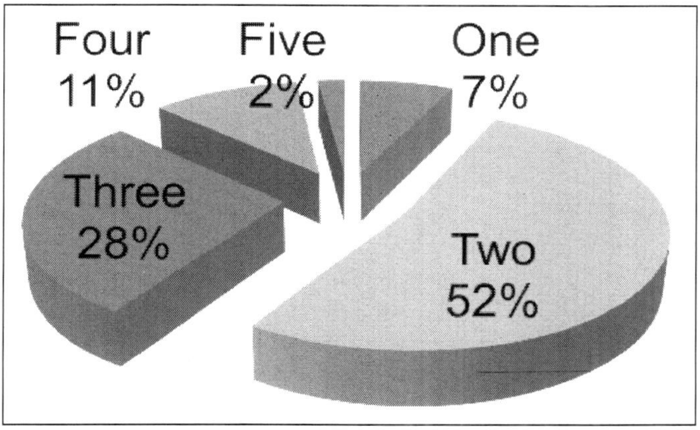

Figure 4.2: Number of Farming System Enterprises Practiced by Marginal Households (Benchmark data of OFR experiment 2 under AICRP on IFS).

Sources of Income Growth

- ✰ Increase in Agricultural Productivity
- ✰ Improvement in total factor productivity
- ✰ Diversification towards high value crops
- ✰ Increase in production of crops
- ✰ Increase in crop intensity
- ✰ Improvement in efficiency of input use
- ✰ Recycling of resources within the system
- ✰ Role of other sectors
    - (i) Improving terms of trade for farmers (MSP *etc.*).
    - (ii) Promotion of subsidiary/complimentary activities

Farming Systems vis-a-vis Management of Natural Resources

The Asian green revolution has shown that if high inputs technology can be effective in terms of productivity, there is a price to pay in terms of environmental degradation, soil infertility, water depletion and contamination, and a loss of biodiversity. In a context where the impact of climate change on the natural resource base is dramatically increasing, adoption of location-specific integrated management of natural resources for higher productivity and better resilience to erratic climatic events is becoming even more crucial. Since small farmers and rural communities are the starting point for efforts to adapt to climate change, the problems and solutions should be defined with their direct and active participation. It is necessary to use participatory farming system processes that empower smallholders to: draw on their expertise about ecology and management; overcome the constraints they face; create a sense of ownership; and share their visions and experiences with other partners. Farming system approach in small and marginal holders are expected to play a vital role in management of natural resources thereby reducing the climatic adversities.

Water Harvesting and Recycling

The production system adopted during green revolution was explorative and the natural resources like soil and water were subjected to immense pressure beyond carrying capacity (Mahapatra *et al.,* 2007). This in turn results in degradation of growing environment and in the absence of integrated management of these vital resources, it will be difficult to achieve the desired level of food production especially rice with threat from the long term changes in weather and micro climate of production system. Farm ponds, as one of the suitable options of land manipulation, form the centre of integrated farming system especially rice based farming system in rainfed environment. Farm ponds may store in-situ rainfall or harvest surface runoff from surrounding areas depending upon the available rainfall in a region. In high rainfall areas, like A and N Islands where average annual rainfall is about 3100 mm, even in-situ rainwater storage in farm pond serves the purpose. However, in areas where surface runoff is the main source of water, the contributing drainage area or watershed should be large enough to maintain desired water level in the farm pond. Following steps should be considered while planning, designing and constructing a farm pond: (i) rainwater availability, (ii) crop water requirements, (iii) design dimension of farm pond, (iv) location of the farm pond and (v) lining requirement for seepage control.

Raised and Sunken Beds

Raised and Sunken Bed (RSB) system also known as Broad Bed and Furrow (BBF) system in Andaman and Nicobar Islands can serve as climate proof technology in the rice based farming systems especially in the coastal areas where in inundation of rice fields are expected due to the sea level rise. It is a technique of land manipulation to grow vegetables, fish and fodder right in the midst of rice fields. The technology involves making of broad bed and furrow alternatively. In the BBF, depressed area is used for rice cultivation and the raised broad bed area, which is above the water level of the paddy field, are used for cultivating of seasonal vegetable or fodder crop during monsoon season. Because of the long term sustainability, easy to adopt and efficient utilization of land area, these techniques is having lot of potential especially for the coastal areas. After through study on dimensions, it was found that beds of 4 m width and furrows of 6 m width with minimum 1 m depth are found suitable for the island conditions having high intensity rainfall. The length of beds and furrows can be according to the length of field. Thus, in one ha area of flat paddy field, 10 beds of 4 m X 100 m X 1 m and 10 furrows of 6 m X 100 m X 1 m can be made which envisages 60 per cent area of furrows and 40 per cent area of beds. The 40 per cent area of beds can be utilized to grow high value vegetables during monsoon season. The broad beds are stabilized by planting two rows of hybrid napier on the edges on either side. Above all this technology is being practiced in rice fields provides best sunshine for growing crops. The system also increases cropping intensity from the present level of 100 per cent in the rice to 300 per cent in the beds and 200 per cent in the furrows of the BBF system. Options of cropping for beds are given in Table 4.1. The raised bed helps to reduce the salinity problem in degraded land and water. Net return of Rs. 1.2 lakhs/year can be obtained from one ha area (Ravisankar *et al.*, 2010).

Table 4.1: Options of Cropping Programme for Beds of BBF

Cropping Sequence	*Period*	*Expected Net Returns (Rs./4000 m^2)*
Option 1		
Cucumber	July -October	17136
Chillies	October - June	192570
Total		**209706**
Option 2		
Amaranthus	July - September	12932
Cabbage	November - February	48394
Amaranthus	February - March	15164
Amaranthus	May - June	24656
Total		**101146**
Option 3		
Bhendi	July - October	12676
Amaranthus	October - November	13916
Coriander	December - February	26068
Bhendi	February - June	12784
Total		**65444**
Option 4		
Crossandra	July - June	20625

Diversification of Small Farms

A farming system model was synthesized using the scientific inputs from the on-station experiment conducted at PDFSR, Modipuram to improve the productivity and profitability of the farm. Accordingly, the cropping systems were modified by including pulses, oilseeds, vegetables, fruits to meet the family demand. The area allocation was also made accordingly. The synthesized cropping systems included sugarcane (spring) + onion-ratoon (12 per cent area, C.12 ha), rice-potato-wheat (0.15 ha)/marigold (0.15 ha)-dhaincha (26 per cent area, 0.30 ha), maize for cobs +arhar-wheat (11 per cent area, 0.13 ha) and sorghum-rice-mustard (0.21 ha)/oat (0.07 ha)/berseem (0.07 ha) (28 per cent area, 0.35 ha). Arhar and mustard were added mainly to produce the sufficient pulses and oilseeds for the family. The livestock component of 2 buffalo and 1 cattle was kept as such but provision for producing sufficient green fodder was kept by including oat and berseem in the cropping systems. In order to enhance, the income and resource recycling, complementary enterprises such as apiary, vermicompost (0.7 per cent area, 100 m2) and karonda, citrus, jackfruit, and subabul as boundary plantation were incorporated. Karonda serves as the live fence and produces fruits which can be used for making pickles. Further, it can also protect the farm from blue bull or stray animals. Mango, Guava + brinjal and tomato (16 per cent area, 0.20 ha) and fishery (7.5 per cent area, 0.08 ha) was added as income supplementing activities in the model.

A 7-member family having 5 adults and 2 children requires 1550, 200, 130, 900, 200, 1120 and 154 kg of cereals, pulses, oilseeds, vegetables, fruits, milk and fish per annum as per ICMR standards to meet the nutritional requirement. It has been found that the synthesized model for the 1.2 ha is able to produce sufficient quantity of these produces required for the family. Apart from this, the system also generates marketable surplus of 3585, 106, 3200, 2218, 5001, 276 kg of cereals, oilseeds, vegetables, fruits, milk and fish respectively ensuring sufficient income for the family besides improving the availability of these products in the market. Additional production of cereals and milk was 265 and 1033 kg/annum respectively. In order to meet the green and dry fodder requirement of 2 buffalos and 1 cattle, around 27 t of green fodder and 5.5 t of dry fodder/annum are required but existing system was producing only 21 t of green fodder. In the improved farming system, the green and dry fodder production increased to 36 and 6.4 t/annum respectively which is the main reason for additional production 1033 kg of milk.

The total production in terms of sugarcane equivalent yield in the improved system was found to be 108 t/annum compared to 56 t/annum only in the existing system. The net profit increases by 88 per cent while the cost of the system increases by 35 per cent only. Internal supply of N, P_2O_5 and K_2O was found to be 204, 136 and 186 kg against only 100, 40 and 100 kg in the existing system. In the improved system, it is estimated that 65, 85 and 100 per cent of N, P_2O_5 and K_2O requirement can be met with in the farm. Further, the recycled resources are expected to supply sufficient level of micro-nutrients. Employment for the family increases from 360 man days to 625 man days.

Bio-intensive Complementary Cropping Systems for Small Farm Holders

Bio-intensive System of raising maize for cobs + vegetable cowpea (1:1 ratio) on tractor made broad beds (BB) and *sesbania* in furrows during *kharif* and mustard in furrows and zero till sown 3 rows of lentil on broad beds in *rabi* while zero till sown 3 rows of green gram on beds in summer (for grain and residue incorporation) is suitable for marginal and small farm holders. The system produces the yield of 18.32 t ha^{-1} as rice equivalent with productivity of 50.2 kg grain $ha^{-1}day^{-1}$ and profitability of Rs.363 $ha^{-1}day^{-1}$. The complimentary effects could be reflected in the system as in broad bed and furrow (BBF) system, the furrows served as drainage channels during heavy rains in *kharif* which were utilized for in-*situ* green manuring with 35 t ha^{-1} green foliage incorporated after 35 days of sowing and timely sown mustard crop in these furrows resulted a good harvest 1.94 t ha^{-1} and a bonus yield of lentil (1.44 t ha^{-1}) can be harvested. In the summer season green gram could yield 1.05 t ha^{-1} grains while incorporation of green foliage of about 4 t ha^{-1} in the soil further helps the system favourably. The cost of cultivation of bio-intensive complementary cropping system ranged from Rs. 48000 to 64500 per ha. The output in terms of rice equivalent yield jumped from 6.7 to 18.3 t ha^{-1}. In overall 40 per cent water, 10-20 per cent energy, 30-40 per cent nutrients and 50 per cent pesticide use could be saved on one hand while productivity could be doubled on the other when compared to existing rice-wheat system (Gangwar and Ravishankar, 2013).

Round the Year Income through Farming System

Rice-poultry-fish-mushroom integration studies conducted during 1987-1992 at Coimbatore revealed that a net profit of Rs. 11,755/year can be obtained in 0.4 ha area while in conventional cropping system with rice-rice-greenmanure/pulses gave a net income of Rs. 6334/year only from the same area (Rangaswamy *et al.*, 1996). The land based enterprises such as dairy, poultry, fishery, mushroom, biogas *etc.* were included by Behera and Mahapatra (1999) to complement the cropping programme to get more income and employment for small farmers of Odisha. A net return of Rs. 58367 can be realized with an investment of Rs. 49286 in 1.25 ha area which also generated 573 man days of employment with a resource use efficiency of Rs. 2.18/Re invested thus ensuring the livelihood of small farmers. Rice based farming system comprising of crop components (Rice-pea-okra and sorghum-berseem-maize), dairy, poultry and fishery was the most suitable and efficient system and recorded higher system productivity and profitability under irrigated ecosystem of eastern Uttar Pradesh (Singh *et al.*, 2006).

Mitigation of GHG Emissions

Recycled paddy straw with mushroom spent substrate was found better over no recycled manure in retaining more moisture (7.70 and 6.69 cm. of soil moisture in top 30 cm. depth, respectively). The depletion of soil moisture was found faster during December–January period as compared to later months. Only recycled poultry manure treatment was able to retain more moisture (4.28 cm.) in the surface layer (0-15 cm depth) compared to sub surface layer (4.18 cm) (Korikanthimath *et al.*, 1998). The pooled mean economic analysis of the integrated systems indicated that rice-brinjal with mushroom and poultry recorded higher gross returns (Rs.1,60,195/ha) as well as net returns (Rs. 77.305/ha) although the cost of production was relatively higher (Rs. 82,890/ha). The system also recorded a moderate benefit-cost ratio (0.93) with better per day net return (Rs. 204/day). Rice-duck farming system is a traditional practice in some of the Asian countries and duck farming is closely associated with wetland rice farming (Panda 2004). The duck integration in rice field increases the rice yield by reducing weed growth, insect population, improving soil physical properties and thereby root growth and tillering. It also increases the dissolved oxygen content in rice field. However, ducks in rice farming reduces the greenhouse effect; prevent the release of methane gas which is important to check global warming. Duck provides additional benefit in terms of egg and meat to the small land holders apart from the environmental benefits.

Strategies for Doubling Farmer's Income

- ☆ Development initiatives
  - (i) Irrigation and Water management
  - (ii) Insurances
  - (iii) Soil health cards
  - (iv) Processing and value addition
- ☆ Technology and Innovations
- ☆ Policies (MSP)
- ☆ Linking Institutions/Stake holders/Markets
- ☆ Road map-IFSA

With changing demand portfolio of various commodities and natural resource availability, the diversification of the farming systems has become imminent in many areas to meet the climate adversities and doubling farmer's income. Efficient utilization of scarce and costly resources in farming system approach is the need of the hour to meet the various climate related adversities. Following the concept of integrated farming systems through supplementation of allied agro-enterprises by recycling the waste of one enterprise in another is a right step which will subsidize the risk of weather vagaries besides providing round the year employment and livelihood. The crisis for shortage of water has already begun in many regions. Better utilization of this scarce resource will bring in more dividends in the long run. Different farming systems modules for small and marginal farmers need to be developed and validated both on-station and on-farm so that the key drivers of high income generation in each of the modules can become the strategic interventions in development programmes across diverse agro-eco conditions for doubling the farmer's income.

References

Gangwar B, Singh J P, Prusty A K and Kamta Pradad 2014 Research in farming systems. Today and Tomorrow's Printers and Publishers, New Delhi pp. 1-584.

Behera, U.K. and I.C. Mahapatra.1999. Income and employment generation of small and marginal farmers through integrated farming systems, *Indian Journal of Agronomy* **44** (3):431-439.

Chand Ramesh 2017. Doubling farmers Income-Rationale, Strategy, Prospects and Action plan NITI Aayog Govt. of India, New Delhi pp. 01-.31.

Korikanthimath, V.S., Mulge, R., Hedge, R., Hiremath, G.G. and Hosmani M.M. 1998. Crop Combination and yield pattern in coffee mix cropped with cardamom. *Journal of Plantation Crops* **26**(1): 41-49

Mahapatra, I.C, J.K. Roy, D.P. Sinhababu and U.K. Behera. 2007. Rice based farming system for livelihood improvement of Indian Farming (In): Extended summaries of National symposium on Research Priorities and Stratagies in rice production system for second green revolution, 20-22 November 2007, Association of Rice Research Workers, CRRI, Cuttack, India pp. 16-18.

Panda, B.K.2004. Rice-duck farming system-A profitable enterprise. In Recent advances in Rice based farming systems, 17-19 November 2004, Panda, D., Sasmal, S, Nayak, S.K., Singh, D.P. and Saha, S (Eds) Central Rice Research Institute, Cuttack, Orissa, pp.202-208.

Rangaswamy, A., R. Venkittaswamy, Purushotaman and S.P. Palaniappan.1996. Rice-poultry-fish-mushroom integrated farming systems for lowlands of Tamil Nadu, *Indian Journal of Agronomy* **41**(3):344-348

Ravisankar, N, T. Subrmani, S.K. Ambast and R.C. Srivastava. 2010. Enhancing farm income in Island ecosystem, *Indian Farming,* **60** (6):16-19.

Singh, K, J.S. Bhora, Y. Singh and J.P. Singh.2006. Development of farming system models for the north eastern plain zone of Uttar Pradesh, *Indian Farming* **56** (2):5-11.

Chapter 5

Enhancing Farm Profitability through Entrepreneurship

Basavaprabhu Jirli

Professor

Department of Extension Education, Institute of Agricultural Sciences

Banaras Hindu University, Varanasi – 221 005, Uttar Pradesh

An overview of literature pertaining to concept of entrepreneur is absolutely essential. The word entrepreneur is derived from **French** word **'Entreprendre'** and the **German** word **"Unternehmen"**, both mean **"To undertake"**. For a long time, there was no equivalent for the term 'entrepreneur' in the English language. Three words were commonly used to connote the sense the French term carried: adventurer, undertaker and projector; these were used interchangeably and lacked the precision and characteristics of a scientific expression (Gopakumar, 1995). Hence the term 'entrepreneur' did not find any prominence in the history of economic thought. The earliest attempt to invest the concept with some economic content could be traced to the works of an 18th century French writer, Bernard F. de Belidor, who defines entrepreneurship as buying labour and materials at uncertain prices and selling the resultant output at contracted prices (Hoselitz, 1960). Entrepreneurship as a concept gathered prominence in economic literature mainly through the writings of Richard Cantillon (1680-1734), who gave the concept some analytical treatment and assigned the entrepreneur an economic role by emphasizing on 'risk' as a prominent entrepreneurial function (Gopakumar, 1995). According to Encyclopaedia Britanica entrepreneur means individual response for the operation of a business, including the choice of a product, the mobilization of necessary capital, decisions on product prices and quantities, the employment of labour and expanding or reducing the productive facilities.

According to Weber (1930), entrepreneurs are a product of the particular social conditions in which they live, and it is the society which shapes the personality of individuals as entrepreneurs. Entrepreneur represents an individual or a group of individuals, who conceive, initiate and maintain for a sufficiently long period of time a social institution which produces economic goods (Cole, 1949) or to put it differently, who perceive a business opportunity and create an organization to pursue it (Bygrave and Hofer, 1991). These definitions help in understanding the concept of entrepreneurship in general, but is there any difference in entrepreneurship in general and in the context of agriculture? The answer is obviously 'yes'.

De (1986) stated that a farmer does not become an entrepreneur only by adopting a new agricultural technology but he becomes an entrepreneur only when he comes to be an operator of a farm business. A business involves rational decisions on investment after assessing risk, other alternatives and possibilities or profit and loss. An entrepreneur is a dynamic agent of change or the catalyst who increasingly transforms the physical, natural, and human resources into corresponding production possibilities.

With this brief background now let's understand whether the Entrepreneurs are born, made and event oriented?

Entrepreneurs are Born

There are certain individuals who are entrepreneurs by birth.

A professor of psychology Jacobwitz, A. came to conclusion after interview with over 500 entrepreneurs over a three-year period that entrepreneurs are commonly share certain personality characteristics *viz.*, 1. Restlessness 2. Independence 3. Tendency to be loner 4. Extreme self confidence.

According to Learned (1992) many individuals are born with "Entrepreneurial DNA" *i.e.* they exhibit a predisposition to entrepreneurship.

Entrepreneurs are Made: According to McClelland (1987) economic activity could be increased through training of business. People can be made entrepreneurs through training of business. He postulated that there is a casual relationship between desire for achievement and business activity.

McClelland (1965) concluded that the need for achievement was essential ingredient for entrepreneurial success. The need for achievement could be thought to stimulate economic growth/entrepreneurial behaviour. We have so many entrepreneurial training institutes for stimulating entrepreneurial behaviour of trainees. Therefore, we can conclude that entrepreneurs are made.

Entrepreneurs are Event Oriented

Many people have become entrepreneurs out of an "Event". According to Knudson *et al.* (2004) event could take many forms including losing one's job, threat of bankrupcy, loss a significant one, frustration on the job, discovering a market place, gap *etc.*

Many times we use the terms entrepreneur and business man synonymously. But there are obvious differences. Understanding these differences is the first step in becoming an entrepreneur. Also along with the concept of entrepreneurship there is concept of social entrepreneurship, there are clear-cut differences between these concepts. Many times we tend to understand the concept of social entrepreneurship as social work, hence these differences are also explained in the further tables and paragraphs.

Differences between Entrepreneurship, Business, Social Entrepreneurship and Non-Governmebntal Organisations

The basic differences between entrepreneurs and businessman include Entrepreneurs create needs while businessmen satisfy needs. Businessmen follow others but entrepreneurs create path (to see the world differently). Entrepreneurs capture opportunities while Businessman unnotice opportunities.

Entrepreneur	*Businessman*
Create Needs	Satisfies Needs
Create Path	Follow others
Capture opportunities	Un-notice opportunities

Differences between Social Entrepreneurship and Entrepreneurship

Social Entrepreneurship	*Entrepreneurship*
Satisfy needs	Creates Needs
Knows exactly what the problem is and struggles with how to create a solution to fix them	Struggles with what the real problems are and build solutions that look for a problem to solve
In social good before they become entrepreneurial first. They understand the social problems better than any other entrepreneur	
Do care about making revenue and turning a profit in a much different way. Revenue is the life blood of any non profit or profit company regardless of social mission	It is very difficult to draw conclusion on social good first and make money secondary or vice-versa
Really entrepreneurs (Truly) with a different mindset	
Improves social conditions	Improves commercial markets
Value lies in social benefit to a community or transformation of a community that lacks resources to fulfil its own needs	Value lies in profit
Creates wealth for effecting social change by changing minds of profit, alleviating suffering and by reversing injustice	Benefits are driven to innovate within a commercial market. The innovation creates wealth.
Older the organisation or initiative the more entrepreneurial skill. Skill are the key to the long term success of the social enterprise	
Less ambitious to grow their employee basis	more ambitious to grow their employee basis
More interested in understanding the social, economic, political and cultural context of the problem they are trying to solve	Not much interested in understanding the social, economic, political and cultural context of the problem they are trying to solve
Delivers social value	Creates economic wealth

Social Entrepreneurship	*Entrepreneurship*
Social returns	Economic returns
Comamnd regards in the society	Demand regards in the society
Mission leader	High achiever
Emotionally charged	Risk bearer
Organiser	Change agent
Opinion Leader	Strategic thinker
Visionery	Arbitrageur (Investor/Trader)
Socially alert	Economically alert

According to Abu-Saifan (2012) common to both qualities include: Innovator, dedicated, initiative, leader, opportunity alert, persistent and committed.

The Differences between Non-Governmental Organisations and Social Entrepreneurship

Non-Governmental Organizations	*Social Entrepreneurship*
Purely Social Purpose	Driven by social as well as financial goal
Source of funding donations, charitable contributions	Earnings from the business
Selection of employees: people participate, voluntary	People are selected based on skill and performance
Performance measurement: on the basis of social values delivered.	on the basis of social values delivered alongwith financial returns.
Runs on social plan for fund raising, social outreach, selling services or products true of cost	Runs on enterprise setting using market tools of commercial and social advertisement for selling products and services
Risk of non-sustainability	Highly sustainable
EX: Helpage India, CRY, AID *etc.*	SKS Microfinance, Rang De, Arvind Eye Care, *etc.*

Having understood the differences between the social work of NGOs and social entrepreneurship let's know the institutions promoting social entrepreneurship.

Examples of Social Entrepreneurship in India

SKS Microfinance

Microfinance is an effective tool that can help reduce poverty and spread economic opportunity by giving poor people access to financial services, such as credit and insurance. SKS distributes small loans that begin at Rs. 2,000 to Rs. 12,000 (about $44-$260) to poor women so they can start and expand simple businesses and increase their incomes. Their micro-enterprises range from raising cows and goats in order to sell their milk, to opening a village tea stall. SKS uses the group lending model where poor women guarantee each other's loans. Borrowers undergo financial literacy training and must pass a test before they are allowed to take out loans. Weekly meetings with borrowers follow a highly disciplined approach.

Re-payment rates on our collateral-free loans are more than 99 per cent because of this systematic process. SKS also offers micro-insurance to the poor as well as financing for other goods and services that can help them combat poverty (http://www.sksindia.com/).

Government of India has an exclusive ministry to promote entrepreneurship and issues associated with. Ministry of Micro, Small and Medium Enterprises is operating with following vision and objectives.

The responsibility of promotion and development of small scale industries (SSI) lies primarily with the State/Union Territory (UT) Governments. However, in recognition of the role of this sector in both creations of wealth and widespread employment and of the need for country-wide framework of policies and measures to facilitate its promotion and development, the Government of India has always supplemented the efforts of the States and UTs in several ways. Over the years, the Central Government has thus formulated policy packages and implemented a number of schemes and programmes, to mainly provide infrastructural, technological, credit, marketing and entrepreneurial development support to the SSI sector.

Vision and Action Plan of the Ministry

The vision of the Ministry of SSI is to create an enabling policy environment and put in place corresponding support measures to help the sector meet the emerging challenges of competition and also harness the opportunities, in tandem with the accelerating pace of liberalization and globalization of the Indian economy. The agenda of the Ministry, therefore, includes steps to develop an appropriate policy framework within which supply of credit is improved, better infrastructural facilities are provided and targeted incentives are designed and extended to the enterprises in the sector for their modernization and technology up gradation, improved market access, entrepreneurship development, skill up gradation and capacity building, both at the level of individual firms and their clusters. The agenda takes cognizance of the significant role played by information technology, sunrise, hi-tech and export potent industries, on the one hand, and, the special needs of the tiny/micro-enterprises, on the other, as they constitute over 95 per cent of the total population of small scale industrial units in the country.

Organisational Set-Up

Most of the programmes and schemes for the development of the sector are implemented through two principal organisations of the Ministry. These are:

i) Small Industries Development Organisation (SIDO); and
ii) National Small Industries Corporation (NSIC).

Besides, there are three national-level Entrepreneurship Development Institutes at Hyderabad, Noida and Guwahati. A National Commission for Enterprises for Unorganised Sector (NCEUS) has also been set up recently by the Government with headquarters in New Delhi.

Small Industries Development Organisation (SIDO)

The major activities of SIDO, established in 1954, include:

- ✰ Advising the Government on policy formulation for promotion and development and enhancement of competitiveness of small-scale industries.
- ✰ Facilitating credit flow to micro and small enterprises through support measures like the Credit Guarantee Fund Scheme.
- ✰ Providing techno-economic and managerial consultancy, common facility and extension services to small-scale units.

The Small Industries Development Organisation (SIDO) comprises the Office of the Development Commissioner (Small Scale Industries) [DC (SSI)] with headquarters in New Delhi, 30 Small Industries Service Institutes (SISIs), 28 Branch SISIs, 4 Regional Testing Centres (RTCs), 7 Field Testing Stations (FTS), one Hand Tool Design Development Training Centre (HTDD and TC), and two Small Entrepreneurs Promotion and Training Institutes (SEPTI) at various locations in the country. It is an Attached Office of the Ministry of SSI. There is at least one SISI located in every State and in some cases more. The 4 RTCs are located at New Delhi, Mumbai, Chennai and Kolkata.

Several grantee institutions/autonomous bodies also function under the administrative control of the SIDO. AS and DC (SSI) is the ex officio Chairman of the Governing Councils of these autonomous societies. The activities of these societies range from research and development to tool design die making, process-cum-product development, training in footwear technology, *etc.*

National Small Industries Corporation Limited (NSIC)

NSIC was established in 1955 with a view to promoting, aiding, and fostering the growth of small scale industries and industry related small scale services/ business enterprises in the country, on a commercial footing. During the last 50 years, the Corporation expanded its activities in the areas of Marketing, Technology, Financing and Information Support to small enterprises in the country. The emerging competitive business environment and dismantling of the market barriers provide an added impetus for enlarging the activities of NSIC. In line with the current economic changes, the Corporation has introduced many new Schemes from time to time for the development of small enterprises. The Corporation is now an ISO: 9001-2000 company, providing services of marketing assistance, technology upgradation and entrepreneurship development. The scheme of assisting these enterprises in obtaining performance – cum – credit rating is the latest initiative of the Government being implemented by the NSIC with the assistance of noted rating agencies operating in the country.

National Institute of Small Industry Extension Training (NISIET), Hyderabad

NISIET was set up in 1960 to assist in promotion, development, and modernisation of small and medium enterprises (SMEs) in the country. At present, the main activities of NISIET are entrepreneurship development related training programmes, training research and consultancy, including the methodology of cluster development.

National Institute for Entrepreneurship and Small Business Development (NIESBUD), Noida

The major activities of the Institute, established in 1986, include, development of model syllabi for training of various target groups, designing effective training strategies, methodology, manuals and tools, facilitating and supporting Central/ State Governments and other agencies in executing programmes of entrepreneurship and small business development, *etc.*

Indian Institute of Entrepreneurship (IIE), Guwahati

IIE was established in 1993, with the aim of undertaking training, research and consultancy activities in the small industry sector focusing on entrepreneurship development as an autonomous national institute. IIE is working towards strengthening the capacity in the field of entrepreneurship development, training, entrepreneurship education, research, and consultancy, incubator services on gems and jewellery, publication and sensitisation of environment for promotion of entrepreneurship, enterprise creation and self-employment in the North Eastern Region. IIE, Guwahati has also taken initiatives for providing hand-holding services to the entrepreneurs in the North Eastern Region for which a Business Facilitation and Development Centre (BFDC) has been set up with financial assistance from the Ministry.

National Commission for Enterprises in the Unorganised Sector

It has been set up by the Government as an advisory body and a watchdog for the informal sector with a view to fulfilling the commitment in the National Common Minimum Programme of the Government. The Commission will recommend measures considered necessary for bringing about improvement in the productivity of these enterprises, generation of large scale employment opportunities on a sustainable basis, particularly in the rural areas, enhancing the competitiveness of the sector in the emerging global environment, linkage of the sector with institutional framework in areas such as credit, raw material, infrastructure, technology upgradation, marketing and formulation of suitable arrangements for skill development.

Apart from these mechanism government brings in various schemes and programs for promoting entrepreneurship among the potential entrepreneurs. The details of efforts, success stories and ways and means to get finance are available in portal developed by Small Industrial Bank of India (SIDBI) you can visit www.smallb.in for details.

Conclusion

Rural development is more than ever before linked to entrepreneurship development. Institutions and individuals promoting rural development now see entrepreneurship as a strategic development intervention that could accelerate the agricultural and rural development process. Furthermore, institutions and individuals seem to agree on the urgent need to promote agricultural enterprises: development agencies see agricultural entrepreneurship as an enormous employment potential; politicians see it as the key strategy to prevent rural unrest; farmers see it as an instrument for improving farm earnings; and women see it as an employment possibility near their homes which provides autonomy, independence and a reduced need for social support. To all these groups, however, entrepreneurship stands as a vehicle to improve the quality of life for individuals, families and communities and to sustain a healthy economy and environment. Entrepreneurship education is often seen as a special kind of training to become an entrepreneur or to enlarge job prospects of the individual. In a broader view entrepreneurial attitudes and skills shall be promoted, in a narrow sense a specific training on how to create a business is meant. An entrepreneur is one who combines the land of one, labor of another and the capital of yet another, and, produces a product. By selling the product in the market, he pays interest on capital, rent on land and wages to laborers. What remains is his or her profit. Around 80 per cent of entrepreneurs fail with their business enterprise. To be successful your product needs to be new to the consumer, and something you think will sell. "At the end we can say that Entrepreneurship is a state of mind if you wish you can be an entrepreneur".

References

Bygrave, W.D. and Hofer, C.F. (1991). Theorizing about entrepreneurship. *Entrepreneurship Theory and Practice*. Vol. (6), No. (2), pp. 13-22.

Cantillon, R. (1955). International encyclopedia of social sciences. Quoted by Cochran, T.C. (1968) under entrepreneurship. MacMillan, New York. Vol. (5), pp. 88-90.

Cole, A.H. (1949). Entrepreneurship and entrepreneurial history. In change and the entrepreneur. Prepared by the research centre in entrepreneurial history. Harward University Press, Cambridge.

Cole, A.H. (1959). Business enterprise in it social setting. Harward University Press, Cambridge.

Cole, A.H. (1968). Meso-economics: A contribution from entrepreneurial history. *Explorations in Entrepreneurial History*. Vol. (6), No. (1), pp. 78-86.

De, D. (1985). Status symbol and innovative entrepreneurship are predictors of farmer's progressivism. *Journal of Extension System*. Vol. (1), No. (1), pp. 81-84.

De, D. (1986). Factors affecting entrepreneur characteristics of farmers. *Indian Journal of Social Work*. Vol. (XLVI), No. (4), pp. 541-546.

De. D and Rao M. S. (2001) Entrepreneurial behaviour of farmers: An Axiomatic theory. Ganga Kaveri Publishing House, D.35/77 Jangamwadi Math Varanasi.

De Dipak, Basavaprabhu Jirli and Gangadharappa N. R. (2017) Entrpreneurship in Agricultural Development. Daya Publishing House, New Delhi (ISBN: 978-93-5124-784-5).

Gopakumar K (1995) Entrepreneurship in economic thought: A thematic review. *Journal of Entrepreneurship*, 4 (1) pp. 1-17.

Hoselitz, B.F. (1957). Non-Economic factors in economic development in Tandoil, B.C. (Eds.). Environmental and entrepreneur. Cheigh Publications, Allahabad, India.

Learned, K.E. (1992). What happened before the organization? A model of organization formation. Entrepreneurship theory and Practice. Vol. 17(1). pp. 39-48.

Mcclelland, D.C. (1953). Achievement motive. Appleton, New York.

Mcclelland, D.C. (1965). Need achievement and entrepreneurship: A longitudinal study. *Journal of Personality and Social Psychology*. Vol. (1), pp. 389-392.

Mcclelland, D.C. (1969). Motivating economic achievement. Free Press, New York.

Chapter 6

Harnessing the Potential of Genomics and Phenomics for Improving Grain Legumes

Chandra Mohan Singh And Mukul Kumar

Department of Genetics and Plant Breeding, Banda University of Agriculture and Technology, Banda – 210 001, Uttar Pradesh

Among the crops, legumes like mungbean, urdbean, cluster bean, French bean, kidney bean, pigeonpea, chickpea, lentil, horse gram, peas, *etc.* are considered as most important, showing a steady increase in terms of production in south and south east Asia. These are cultivated throughout year *like* kharif (rainy), rabi (winter) and zaid (spring/summer) season in different parts of the country. These crops have valuable nutritional and health benefits, especially in developing countries, where malnutrition is a major issue. It represents a cheap source of carbohydrates and vegetarian protein. It is a good substitute for non-vegetarian protein in most Asian diet and an excellent source of essential nutrients and also improves the soil health by fixing the atmospheric nitrogen. The certain features make it a well-suited model plant among legumes because of its small genome size, short life-cycle, self-pollinating, and close genetic relationship to other legumes. But various factors including non-availability of high yielding varieties suited to erratic climatic condition *etc.*, are responsible for poor yield.

Now the days among the various abiotic stresses, heat and drought stresses are major concern which is highly injurious and affects different phenophases leading to severe yield loss It causes many undesirable effects *like* cell death, burning, flower drop, pollen abortion and shortening the grain filling duration, looses in stigma receptivity and ovule viability. These are polygenic trait which adversely affected the morpho-physiological and biochemical characters. Adoption of mungbean and genotypes in summer season may improve the livelyhood security of the farmers coupled with nutritional and soil health benefit. Deployment of high teperature tolerant mungbean and urdbean genotypes may be most effective, economic and eco-friendly way to adress the issue. In the past, modest efforts have been made to develop molecular markers associated with genetic improvement of mungbean and urdbean. Lack of a reliable parameter for screening of heat tolerance and non-avaialability of linked markers make it more challenging task to develop heat tolerance cultivars. Due to the complex nature of stress, little progress has been made in identifying heat tolerant mungbean and urdbean genotypes. Similarly, biotic stresses like yellow mosaic disease, cercospora leaf spot, web blight, powdery mildew, white fly, hairy caterpillar, thrips, *etc.* are major concern for mungbean and other *Vigna* production. Likewise, stemphylium, wilt, pod borer is major concern of pigenpea, chickpea and lentil.

Molecular marker technology increases the efficiency of breeding through marker assisted selection (MAS), in which phenotypic selection is carried out using DNA markers associated with the trait of interest, which may accelerate the development of modern cultivars. The conventional breeding approach coupled with molecular marker technology has proven to be effective in introgressing resistance in legumes (Varshney *et al.*, 2015; Pratap *et al.*, 2017), which can help in accelerating mungbean improvement programmes also. In mungbean molecular breeding now sounds more practicable after the availability of whole genome sequence of different grain legumes. In recent, transcriptome profiling is in more practice for tagging differentially expressed genes and developing gene specific markers/EST-markers associated with drought/heat tolerance is one of the effective approach, which can be effectively utilized in fine-mapping of the target locus and marker-assisted selection. But still, breeders is not able to utilize the all genomic resources due to several constraints, in which phenotyping is major issue. In such situation high throughput phenotyping give more sound data which might be linked with the genomic and transcriptomic data for understanding the machanism of different resistance and help in marker-assisted breeding of grain legumes. Plant phenotyping techniques have evolved impressively during last two decades and low-cost, automated and semi-automated methods for data acquisition, storage and analysis have been developed which allow a precise quantitative analysis of plant structure and function and genetic dissection of complex traits (Pratap *et al.*, 2019). A various high throughput phenotyping platforms are available now the days like PHENOPSIS, WIWAM, PlantScan, PlantScreen™, Rosette Tracker, Leaf Analyser, *etc.* for proper screening of the traits. The phenotyping databases along with available genomic databases have made it more feasible for dissecting the genetic architecture of several complex traits (Mir *et al.*, 2015). Both forward and reverse phenomics approaches can be used to harness the potentiality of genomic resources (Furbank and Tester, 2011). By linking the omics tools with phenotypic data and their effective utilization will provide a means to improvement in the grain legumes.

References

Furbank RT, Tester M. Phenomics – technologies to relieve the phenotyping bottleneck. Trends Plant Sci. 2011;16: 635–644.

Mir RR, Choudhary N, Singh B, Khandy IA, Bawa V, Sofi P, Wani A, Kumari S, Jain S, Kumar A. Harnessing Genomics Through Phenomics. In Phenomics in Crop Plants: Trends, Options and Limitations. Springer, New Delhi. 2015. pp. 273-283).

Pratap, A., Chaturvedi, S. K., Tomar, R., Rajan, N., Malviya, N., Thudi, M., Saabale, P. R., Prajapati, U., Varshney, R. K. and Singh, N. P. (2017). Marker-assisted introgression of resistance to fusarium wilt race 2 in Pusa 256, an elite cultivar of *desi* chickpea. *Molecular Genetics Genomics,* 292, 1237-1245

Pratap, A.; Gupta, S.; Nair, R.M.; Gupta, S.K.; Schafleitner, R.; Basu, P.S.; Singh, C.M.; Prajapati, U.; Gupta, A.K.; Nayyar, H.; Mishra, A.K.; Baek, K.-H. Using Plant Phenomics to Exploit the Gains of Genomics. *Agronomy* **2019**, *9*, 126. https://doi.org/10.3390/agronomy9030126.

Varshney, R. K., Mohan, S. M., Gaur, P. M., Chamarthi, S. K., Singh, V. K., Srinivasan, S., Swapna, N., Sharma, M., Pande, S., Singh, S. and Kaur, L. (2014). Marker-assisted backcrossing to introgress resistance to Fusarium wilt race 1 and Ascochyta blight in C 214, an elite cultivar of chickpea. *The Plant Genome,* 7, 1–11.

Chapter 7

Wealth from Agricultural Waste

Dinesh Sah and Aniket H. Kalhapure

Department of Agronomy, College of Agriculture,
Banda University of Agriculture and Technology, Banda, Uttar Pradesh

Current trends in world population growth indicate that the demand for food will steadily increase, creating greater amounts of agricultural by-products. Each year, approximately one billion tonnes of these by-products are created through the production of food worldwide. These by-products are unavoidable and were originally considered to be largely unusable. Finding ways to dispose of the waste with minimal impact on the environment is one of today's greatest challenges. **Waste** Recycling, thus, has been one major and significant step in the sector of environment as well as to develop entrepreneurship based solutions to such critical problems. The concept of **Waste-to-Wealth** literally **means** moving **waste** from a platform of exhausted utility to valuable and desirable level.

Composting agricultural and kitchen food waste is easy and requires little time, effort or space, depending on which system you use. The compost is invaluable for the soil in your farm, garden or potted plants. It's a complete and natural food for the soil, helping to improve its structure, water-retaining abilities and overall health. Composting is an aerobic, biological process that uses naturally occurring microorganisms to convert biodegradable organic matter into humus like product. The process destroys pathogens, converts nitrogen from unstable ammonia to stable organic form and reduces the volume of the waste. Presently composting has become an important method to dispose of farm wastes.

Aerobic composting is a technology of quick disposal of garbage and desirable recycling process because considerable portion of farm waste is biodegradable material and with high moisture content. Much of the research on composting has focused on the changes in the physical and chemical parameters of the compost primarily in an effort to find a simply and reliable indicator of compost maturity and to improve the efficiency of the process. Compost is a rich source of organic matter. Soil organic matter plays an important role in sustaining soil fertility, and hence in sustainable agricultural production. In addition to being a source of plant nutrient, it improves the physico-chemical and biological properties of the soil. As a result of these improvements, the soil:

1. Becomes more resistant to stresses such as drought, diseases and toxicity;
2. Helps the crop in improved uptake of plant nutrients; and
3. Possesses an active nutrient cycling capacity because of vigorous microbial activity. These advantages manifest themselves in reduced cropping risks, higher yields and lower outlays on inorganic fertilizers for farmers.

Why Composting is Necessary?

- ☆ The rejected biological materials contain complex chemical compounds such as lignin, cellulose, hemicellulose, polysaccharides, proteins, lipids *etc.*
- ☆ These complex materials cannot be used as such as resource materials.
- ☆ The complex materials should be converted into simple inorganic element as available nutrient.
- ☆ The material put into soil without conversion will undergo conversion inside the soil.
- ☆ This conversion process takes away all energy and available nutrients from the soil affecting the crop. Hence conversion period is mandatory.

Dung and Urine Produced by Animals per day

Animal	*Urine (ml/kg live wt)*	*Quantity of Dung (Kg) per Day*
Horse	3-18	9-18
Cattle	17-45	18-30
Buffaloes	20-45	25-40
Sheep and goats	10-40	1-2.5
Pigs	5-30	3-5
Poultry	-	2.5-3.5

Nutrient Value of Animal Solid and Liquid Excreta

Animal	*Dung (mg/g)*			*Urine (per cent)*		
	N	*P*	*K*	*N*	*P*	*K*
Cattle	20-45	4-10	7-25	1.21	0.01	1.35
Sheep and goat	20-45	4-11	20-29	1.47	0.05	1.96
Pig	20-45	6-12	15-48	0.38	0.1	0.99
Poultry	28-62	9-26	8-29	-	-	-

Advantages of Composting

- Volume reduction of waste.
- Final weight of compost is very less.
- Composting temperature kill pathogen and weed seeds.
- Matured compost comes into equilibrium with the soil.
- During composting number of wastes from several sources are blended together.
- Excellent soil conditioner
- Saleable product
- Improves manure handling
- Redues the risk of pollution
- Pathogen reduction
- Additional revenue.
- Suppress plant diseases and pests.
- Reduce or eliminate the need for chemical fertilizers.
- Promote higher yields of agricultural crops.
- Facilitate reforestation, wetlands restoration, and habitat revitalization efforts by amending contaminated, compacted, and marginal soils.
- Cost-effectively remediate soils contaminated by hazardous waste.
- Remove solids, oil, grease, and heavy metals from stormwater runoff.
- Capture and destroy industrial volatile organic chemicals (VOCs) in contaminated air.
- Provide cost savings of at least 50 per cent over conventional soil, water, and air pollution remediation technologies, where applicable.

Limitations of Using Composts

Agricultural use of composts remains low for several reasons:

- The product is weighty and bulky, making it expensive to transport.
- The nutrient value of compost is low compared with that of chemical fertilizers, and the rate of nutrient release is slow so that it cannot usually meet the nutrient requirement of crops in a short time, thus resulting in some nutrient deficiency
- The nutrient composition of compost is highly variable compared to chemical fertilizers.
- Agricultural users might have concerns regarding potential levels of heavy metals and other possible contaminants in compost, particularly mixed municipal solid wastes. The potential for contamination becomes an important issue when compost is used on food crops.

- ☆ Long-term and/or heavy application of composts to agricultural soils has been found to result in salt, nutrient, or heavy metal accumulation and may adversely affect plant growth, soil organisms, water quality, and animal and human health.

The Benefits of Using Composts to Agriculture

Compost has been considered as a valuable soil amendment for centuries. Most people are aware that using composts is an effective way to increase healthy plant production, help to save money, reduce the use of chemical fertilizers, and conserve natural resources. Compost provides a stable organic matter that improves the physical, chemical, and biological properties of soils, thereby enhancing soil quality and crop production. When correctly applied, compost has the following beneficial effects on soil properties, thus creating suitable conditions for root development and consequently promoting higher yield and higher quality of crops.

Improves the Physical Properties of Soils

- ☆ Reduces the soil bulk density and improves the soil structure directly by loosening heavy soils with organic matter, and indirectly by means of aggregate-stabilizing humus contained in composts. Incorporating composts into compacted soils improves root penetration and turf establishment.
- ☆ Increases the water-holding capacity of the soil directly by binding water to organic matter, and indirectly by improving the soil structure, thus improving the absorption and movement of water into the soil. Therefore, water requirement and irrigation will be reduced.
- ☆ Protects the surface soil from water and wind erosion by reducing the soil-dispersion action of beating raindrops, increasing infiltration, reducing water runoff, and increasing surface wetness. Preventing erosion is essential for protecting waterways and maintaining the quality and productivity of the soil.
- ☆ Helps bind the soil particles into crumbs by the fungi or actinomycetes mycelia contained in the compost and stimulated in the soil by its application, generally increasing the stability of the soil against wind and water erosion.
- ☆ Improves soil aeration and thus supplies enough oxygen to the roots and escapes excess carbon dioxide from the root space.
- ☆ Increases the soil temperature directly by its dark color, which increases heat absorption by the soil, and indirectly by the improved soil structure.
- ☆ Helps moderate soil temperature and prevents rapid fluctuations of soil temperature, hence, providing a better environment for root growth. This is especially true of compost used as a surface mulch.

Enhances the Chemical Properties of Soils

- ☆ Enables soils to hold more plant nutrients and increases the cation exchange capacity (CEC), anion exchange capacity (AEC), and buffering capacity of soils for longer periods of time after composts are applied to soils. This is important mainly for soils containing little clay and organic matter.
- ☆ Builds up nutrients in the soil. Composts contain the major nutrients required by all plants [N,P,K, calcium (Ca), magnesium (Mg), and S] plus essential micro-nutrients or trace elements, such as copper (Cu), zinc (Zn), iron (Fe), manganese (Mn), boron (B), and molybdenum (Mb).
- ☆ The nutrients from mature composts are released to the plants slowly and steadily. The benefits will last for more than one season.
- ☆ Stabilizes the volatile nitrogen of raw materials into large protein particles during composting, thereby reducing N losses.
- ☆ Provides active agents, such as growth substances, which may be beneficial mainly to germinating plants.
- ☆ Adds organic matter and humus to regenerate poor soils.
- ☆ Buffers the soil against rapid changes due to acidity, alkalinity, salinity, pesticides, and toxic heavy metals.

Improves the Biological Properties of Soils

- ☆ Supplies food and encourages the growth of beneficial microorganisms and earthworms.
- ☆ Helps suppress certain plant diseases, soil borne diseases, and parasites.
- ☆ Research has shown that composts can help control plant diseases (*e.g.* Pythium root rot, Rhizoctonia root rot, chili wilt, and parasitic nematode) and reduce crop losses. Research has also indicated that some composts, particularly those prepared from tree barks, release chemicals that inhibit some plant pathogens. Disease control with compost has been attributed to four possible mechanisms:
  1. Successful competition for nutrients by beneficial microorganisms;
  2. Antibiotic production by beneficial microorganisms;
  3. Successful predation against pathogens by beneficial microorganisms;
  4. Activation of disease-resistant genes in plants by composts; and
  5. High temperatures that result from composting kill pathogens.
- ☆ Reduces and kills weed seeds by a combination of factors including the heat of the compost pile, rotting, and premature germination.

Economic and Social Benefits of Composting

The economic and social benefits of composting include the following:

- Brings higher prices for organically grown crops.
- Composting can offer several potential economic benefits to communities:
- Extends current landfill longevity and delays the construction of a more expensive replacement landfill or incinerator.
- Reduces or avoids landfill or combustor tipping fees, and reduces waste disposal fees and long-distance transportation costs.
- Offers environmental benefits from reduced landfill and combustion use.
- Creates new jobs for citizens.
- Produces marketable products and a less-cost alternative to standard landfill cover, artificial soil amendments, and conventional bioremediation techniques.
- Provides a source of plant nutrients and improves soil fertility; results in significant cost savings by reducing the need for water, pesticides, fungicides, herbicides, and nematodes.
- Used as an alternative to natural topsoil in new construction, landscape renovations, and container gardens. Using composts in these types of applications is not only less expensive than purchasing topsoil, but it can also often produce better results when establishing a healthy vegetative cover.

Used as mulch for trees, orchards, landscapes, lawns, gardens, and makes an excellent potting mix. Placed over the roots of plants, compost mulch conserves water and stabilizes soil temperatures. In addition, it keeps plants healthy by controlling weeds, providing a slow release of nutrients, and preventing soil loss through erosion.

Composting Organic Materials with High Lignin Content - Lime Treatment

- By adding organic wastes such as sawdust, wood shavings, coir pith, pine needles, and dry fallen leaves, while preparing organic waste mixtures for composting, one can ensure that the compost produced contains sufficient and long-lasting humus. A technique for making good compost from hard plant materials involves mixing lime in a ratio of 5 kg per 1000 kg of waste material. Lime can be applied as dry powder or after mixing with a sufficient quantity of water. Treatment with lime enhances the process of decomposition of hard materials.
- Liming can enhance the humification process in plant residues by enhancing microbial population and activity and by weakening lignin structure. It also improves the humus quality by changing the ratio of humic to fulvic acids and decreases the amount of bitumen, which interferes with the decomposition process. Instead of lime, powdered rock

phosphate can be used in a ratio of 20 kg per 1000 kg of organic waste. Phosphate rock contains a lot of lime. The phosphates and micro-nutrients contained in phosphate rock make composts rich in plant nutrients.

1. Composting Weeds

This method has been developed for composting weeds such as parthenium, water hyacinth (*Eichornia crassipes*), cyperus (*Cyperus rotundus*) and cynodon (*Cynodon dactylon*).

1.1 Materials Required

250 g of *Trichoderma viride* and *Pleurotus sajor-caju* consortia, and 5 kg of urea. An elevated shaded place is selected, or a thatched shed is erected. An area of 500 cm × 150 cm is marked out. The material to be composted is cut to 10-15 cm in size. About 100 kg of cut material is spread over the marked area. About 50 g of microbial consortia is sprinkled over this layer. About 100 kg of weeds are spread on this layer. One kilogram of urea is sprinkled uniformly over the layer. This process is repeated until the level rises to 1 m. Water is sprinkled as necessary to maintain a moisture level of 50-60 per cent. Thereafter, the surface of the heap is covered with a thin layer of soil. The pile requires a thorough turning on the twenty-first day. The compost is ready in about 40 days.

1.2 Compost Enrichment

Farm compost is poor in P content (0.4-0.8 per cent). Addition of P makes the compost more balanced, and supplies nutrient to micro-organisms for their multiplication and faster decomposition. The addition of P also reduces N losses. Compost can be enriched by:

- Application of superphosphate, bone meal or phosphate rock: 1 kg of superphosphate or bone meal is applied over each layer of animal dung. Low-grade phosphate rock can also be used for this purpose.
- Use of animal bones: these can be broken into small pieces, boiled with wood ash leachate or lime water and drained, and the residue applied to the pits. This procedure of boiling bones facilitates their disintegration. Even the addition of raw bones, broken into small pieces and added to the pit, improves the nutrient value of compost significantly.
- Wood ash waste can also be added to increase the K content of compost.
- Addition of N-fixing and P-solubilizing cultures (IARI, 1989): The quality of compost can be further improved by the secondary inoculation of *Azotobacter*, *Azospirillum lipoferum*, and *Azospirillum brasilence* (N-fixers); and *Bacillus megaterium* or *Pseudomonas* sp. (P solubilizers). These organisms, in the form of culture broth or water suspension of biofertilizer products, can be sprinkled when the decomposing material is turned after one month. By this time, the temperature of the compost has also stabilized at about 35 °C. As a result of this inoculation, the N content of straw compost can be increased by up to 2 per cent. In addition to

improving N content and the availability of other plant nutrients, these additions help to reduce the composting time considerably.

2. Crop Residue Composting

Crop residues are the non-economic plant parts that are left in the field after harvest. The harvest refuses include straws, stubble, stover and haulms of different crops. Crop remains are also from thrashing sheds or that are discarded during crop processing. This includes process wastes like groundnut shell, oil cakes, rice husks and cobs of maize and sorghum. The greatest potential as a biomass resource appears to be from the field residues of sorghum, maize, soybean, cotton, sugarcane *etc.*

2.1 Waste Collection

Crop residues accumulated in different locations are to be brought to compost yard. The compost yard is located in anyone corner of the farm with accessibility via good road. Water resource should also be available in sufficient quantity. The crop residues that are brought to compost yard should be heaped in one corner for further processing.

2.2 Shredding of Waste Materials

Particle size is one of the factors that influence the composting. It is advisable to shred all the crop residues that are used for composting. Shredding the waste manually is labour intensive. Shredder machine can be employed to shred all the crop residue biomass. Particle size of 2 to 2.5 cm is recommended for quick composting.

2.3 Mixing of Green Waste and Brown Waste

Carbon and nitrogen ratio decides the initiation of composting process. If C:N ratio is wide (100:1) composting will not take place. Narrow C:N ratio of 30:1 is ideal for composting. To get a narrow C:N ratio, carbon and nitrogen rich material should be mixed together. Green coloured waste materials like glyricidia leaves, parthenium, freshly harvested weeds; sesbania leaves are rich in nitrogen, whereas brown coloured waste material like straw, coir dust, dried leaves and dried grasses are rich in carbon. In any composting process these carbon and nitrogen rich material is to be mixed together to make the composting quicker rather than putting green waste alone or brown color waste alone for composting. Animal dung is also a good source of nitrogen. While making heap formation, alternative layers of carbon rich material, animal dung and nitrogen rich material are to be heaped to get a quicker result in composting.

2.4 Compost Heap Formation

Minimum 4 feet height should be maintained for composting. The composting area should be elevated one and have sufficient shade. While heap formation, all the crop residues should be mixed together to form a heterogeneous material rather than a single homogenous material. Alternate layers of carbon and nitrogen rich material with intermittent layers of animal dung are essential. After heap formation the material should be thoroughly moistened.

2.5 Bio Inputs for Composting

NCOF compost decomposer contains groups of microorganisms, which accelerate the composting process. If this inoculum is not added to the composting material, natural microorganisms establish on the waste material on its own and do the composting work. This process takes longtime. But if external source of inoculums is added, the microbial activity starts earlier and composting period will be reduced.

For one ton of crop wastes a small quantity of Biomineralizer is recommended. This two kg Biomineralizer should be mixed with 20 liters of water and made slurry. When the compost heap is formed in between layers the slurry should be inoculated, so that it mixes with the waste material thoroughly for uniform coating of microorganism on the waste material. Cow dung slurry is also a good source for microbial inoculum. But it carries unwanted microorganisms also which may compete with composting organism. But when Biomineralizer is not available, cow dung slurry is a good source material. For one ton crop residues40 kg fresh cow dung is required. This 40 kg fresh cow dung is mixed with 100litres of water and it should be thoroughly poured over the waste material. Cow dung slurry acts as nitrogen source as well as source of microbial inoculum.

2.6 Aerating the Compost Material

Sufficient quantity of oxygen should be available inside the compost heap. For this external air should be freely get in and comes out of the material. Normally to allow the fresh air to get inside, the compost heap should be turned upside down, once in fifteen days. In this process top layer comes to bottom and bottom layer goes to top. This process also activates the microbial process and compost process is hastened. In some cases air ventilating pipe maybe inserted vertically and horizontally, to allow the air to pass through. The wood chip that is available as waste in wood processing industry may also be used as bulking agent in the composting process. This bulking agent gives more air space to the compost material.

2.7 Moisture Maintenance

Throughout the composting period 60 per cent moisture should be maintained. On any situation, compost material should not be allowed to dry. If the material becomes dry, all the microorganisms present in the crop residues will die and the compost process gets affected.

2.8 Compost Maturity

Volume reduction, black colour, earthy odor, reduction in particle size are all the physical factors to be observed for compost maturity. After satisfying with the compost maturity index, the compost heap can be disturbed and spread on the floor for curing. After curing for one day, the composted material is sieved through 4 mm sieve to get uniform composted material. The residues collected after composting has to be again composted to finish the composting process.

2.9 Compost Enrichment

The harvested compost should be heaped in a shade, preferably on a hard floor. The beneficial microorganisms like *Azotobacter* or *Azospirillum, Pseudomonas,* Phosphobacteria (0.2 per cent) and rock phosphate (2 per cent) have to be inoculated for one ton of compost. Forty per cent moisture should be maintained for the maximum growth of inoculated microorganism. This incubation should be allowed for 20 days for the organism to reach the maximum population. Now the compost is called as enriched compost. The advantage of enriched compost over normal compost is the quality manure with higher nutrient status with high number of beneficial microorganisms and plant growth promoting substances.

The common manurial value of different bio-compost prepared from different substrate is given below.

Nutrient Content of Bio-compost Prepared from different Crop Residues

Biocompost	*Nutrient Content (per cent)*		
	Nitrogen	*Phosphorous*	*Potash*
Animal refuse			
Cattle dung	0.3 - 0.4	0.1 – 0.2	0.1 – 0.3
Horse dung	0.4 – 0.5	0.3 – 0.4	0.3 – 0.4
Sheep dung	0.5 – 0.7	0.4 – 0.6	0.3 - 0.1
Poultry manure	1.8 – 2.2	1.4 – 1.8	0.8 – 0.9
Sewage sludge	2.0 – 3.5	---	---
Cattle urine	0.9 – 1.2	Trace	0.5 – 1.0
Horse urine	1.2 – 1.5	Trace	1.3 – 1.5
Sheep urine	1.5 – 1.7	0.1 – 0.2	0.1 – 0.3
Wood Ash			
Ash coal	0.73	0.45	0.53
Ash wood	0.1 – 0.2	0.8 – 5.9	1.5 – 36.00
Habitation waste and factory waste			
Rural compost	0.5 – 1.0	0.4 – 0.8	0.8 – 1.2
Urban compost	0.7 – 2.0	0.9 – 3.0	0.3 – 1.9
Farmyard manure	0.4 – 1.5	0.3 – 0.9	2.0 – 7.0
Straw and stalk			
Pearl millet	0.65	0.75	2.50
Cotton	0.44	0.10	0.66
Banana pseudo stem	0.61	0.12	1.00
Sorghum	0.40	0.23	2.17
Maize	0.42	1.57	1.65
Paddy straw	0.36	0.08	0.71
Tobacco	1.12	0.84	0.80
Pigeon pea	1.10	0.58	1.28

Biocompost	Nutrient Content (per cent)		
	Nitrogen	Phosphorous	Potash
Sugarcane trash	0.53	0.10	1.10
Wheat	0.53	0.10	1.10
Tobacco dust	1.10	0.31	0.93

Compost Application

Organic manures are highly regarded as good source of material to maintain soil health and increasing soil organic carbon content. Organic manures cannot be equated with inorganic fertilizers. But organic manures deliver all the nutrients to the soil but with little quantity. For one hectare of land 5 tons of enriched biocompost is recommended. It can be used as basal application in the field before taking up sowing or planting work.

3. Technology for Composting of Poultry Wastes

Preparation of Poultry Waste Compost using Paddy Straw

Inputs Required

- Poultry droppings
- Paddy straw
- Pleurotus sajor-caju

A known quantity of fresh poultry droppings is to be collected and mixed thoroughly with chopped paddy straw (< 2 cm size) @ 1:1.25 ratio so as to attain a C/N ratio of 25 to 30 which is considered to be optimum for composting. *Pleurotus sajor-caju* is inoculated @ 5 packets (250 g each) per tonne of substrate. The poultry waste and paddy straw mix should be heaped under shade. The moisture content of the heap should be maintained at 50 to 60 per cent. Periodical watering should be done once in 15 days and turning should be given on 21st, 35th and 42nd day of composting (avoid turning during first 3 weeks of composting). Within a period of 50 days, materials are converted to matured compost with the following nutrient contents;

- N: 1.89 per cent
- P: 1.83 per cent
- K: 1.34 per cent
- C/N: 12.20

References

Mahimairaja,S., P.Dooraisamy, A.Lakshmanan, G.Rajannan, C.Udayasoorian and S.Natarajan.2008.Composting technology and organic waste utilization in Agriculture.A.E.Publications, P.N.Pudur, Coimbatore=641 041

www.fao.org

Chapter 8

Salt Affected Soils and their Management

Jagannath Pathak[1], A.K. Chaubey[1], J.K. Tiwari[1], Amit Mishra[1], Rajeev Shukla[2], Anoop Dixit[3] and B.P. Mishra[1]

[1]College of Agriculture, BUAT, Banda – 210 001 (UP)
[2]GB Pant University of Agriculture and Technology, Pantnagar
[3]ICAR-Indian Grassland and Fodder Research Institute, Jhansi (UP)

Introduction

All soils contain some amounts of soluble salts. Many of these salts act as a source of essential nutrients for the healthy growth of plants. However, when the quantity of the salts in the soil exceeds a particular limit, the growth, yields and/or quality of most crops is adversely affected depending upon the kind and amount of salt present. Soil that contains excess amount of soluble salts such as chlorides, sulphates, carbonates and bicarbonates of calcium, magnesium and sodium so as to impair the productivity of crops is called salt affected soil.

Salinity as a prominent form of land degradation affects over 1100 m ha lands worldwide (Wicke *et al.*, 2011). Although predominant in arid and semi-arid regions, saline and sodic soils occur in almost all climates. Besides high salt concentrations in root zone, saline and sodic soils are also mostly underlain with marginal quality waters (Singh, 2009). Both of these problems are likely to significantly increase in future.

Estimates of Salt Affected Soils

According to the FAO Land and Plant Nutrition Management Service, over 6 per cent of the world's land is affected by either salinity or sodicity. The term *salt- affected* refers to soils that are saline or sodic, and these cover over 400 million hectares, which is over 6 per cent of the world land area. Much of the world's land is not cultivated, but a significant proportion of cultivated land is salt-affected. Of the current 230 million ha of irrigated land, 45 million ha are salt-affected (19.5 per cent). In India, about 6.73 M ha of land is affected by salinity and sodicity problems.

Soil Resources of India

India's share in land resources of the world is only 2 per cent, on which 18 per cent of the world's population and over 15 per cent of the world's livestock survive. However, with its diverse agro-climate, topography and soil types, India is capable of producing a wide range of crops and vegetation. The land surface of the country is spread over an area of 329 M ha and is represented by different types of soils which are given in the (Table 8.1).

Table 8.1: Major Soil Groups in India

Soils	*Area (m.ha)*
Red and laterite soils	117.2
Black soils	73.5
Alluvial soils	58.4
Desert soils	30.0
Other soils (Saline alkali soils, Forest and hill soils, Peaty and Marshy Soils)	49.6

Soil Degradation

The primary cause of degradation is the demographic pressure on land, resulting in loss of vegetal cover through deforestation. The land degradation occurs mainly due to uncontrolled deforestation followed by agricultural/farm activities. Hence, planning for productive land use is necessary to meet the growing challenges of food security since the land resource is not expandable physically.

Salt Affected Soils

In India salt affected soils are mainly confined to the arid and semi-arid and subhumid (dry) regions and also in the coastal areas. The salt deposits are of sodium carbonate, sulphate and chloride with calcium and magnesium.

- ✰ These soils vary in nature from saline to non-saline sodic.
- ✰ In coastal regions saline soils are most predominant. They have high soluble salts (EC >4 dS/m) of chloride and sulphate of sodium, calcium and magnesium, low ESP and have pH value less than 8.2.

Extent of Salt Affected Soils in India

The salt affected soils were classified according to norms for pH, electrical conductivity (EC) and exchangeable sodium percentage (ESP). The state wise extent of salt affected soils in India is given in Table 8.2. It shows that maximum area of salt affected soils occur in Gujarat followed by Uttar Pradesh and Maharashtra which account for about 62.4 per cent. Due to the limitation of small scale some very small and isolated patches of salt affected soils occurring in the states of Delhi and Himachal Pradesh could not be detected. The salt affected soils accounts for 6.727 m ha equivalent to 2.1 per cent of the geographical area of the country. Out of the total 6.727 million ha of salt affected soils, 2.956 million ha are saline and the rest 3.771 million ha are sodic. Out of the total 2.347 million ha salt affected soils in the Indo-Gangetic Plains, 0.56 million ha are saline and 1.787 million ha are sodic.

Table 8.2: Extent of Salt-Affected Soils in India ('000 ha)

State	*Saline*	*Sodic*	*Total*
Andhra Pradesh	77.598	196.609	274.207
Andaman and Nicobar Island	77.000	0	77.000
Bihar	47.301	105.852	153.153
Gujarat	1680.570	541.430	2222.000
Haryana	49.157	183.399	232.556
Karnataka	1.893	148.136	150.029
Kerala	20.000	0	20.000
Madhya Pradesh	0	139.720	139.720
Maharashtra	184.089	422.670	606.759
Orissa	147.138	0	147.138
Punjab	0	151.717	151.717
Rajasthan	195.571	179.371	374.942
Tamil Nadu	13.231	354.784	368.015
Uttar Pradesh	21.989	1346.971	1368.960
West Bengal	441.272	0	441.272
Total	2956.809	3770.659	6727.468

Source: NRSA and Associates (1996).

Characteristics of Salt Affected Soils

The term "soil" is considered to be a three-dimensional piece of landscape having shape, area and depth. Saline and alkali soils are defined and diagnosed on the basis of EC and SAR determination made on soil samples and the information thus generated contributes substantially to the scientific agriculture based on USDA classification is given in Table 8.3.

Saline Soils

Saline soils contain neutral soluble salts of chlorides and sulfates of sodium, calcium and magnesium. The electrical conductivity of saturated extract of the soil is more than 4 dS m^{-1}. ESP is less than 15 and pH is less than 8.5. Because of the presence of excess salts and low amount of Na+, these soils are in a flocculated state and their permeability is higher than alkali soils. Their physical condition is good and water can pass through them. These soils have a white crust of salts on their surface. Salinization refers to the accumulation of neutral soluble salts in soils.

Table 8.3: Properties of Saline, Saline-Alkali and Non-Saline-Alkali Soils

Properties	*Saline Soils*	*Saline Alkali Soils*	*Non-saline Alkali Soils*
Electrical conductivity (dS m^{-1})	> 4.0	> 4.0	< 4.0
pH	< 8.5	> 8.5	> 8.5
Exchangeable Sodium Per cent	< 15	> 15	> 15

Formation of Saline Soils

Saline soils are formed whenever climate, soil and hydrological conditions favour accumulation of soluble salts in the root zone. Arid and semi-arid regions of low rain fall and high temperature leads to less leaching, more evaporation of water and less transport of soluble salts. Irrigation with salt-laden underground water and their indiscriminate use under inadequate drainage creates more salinity problems. Excessive use of chemical fertilizers may also lead to form saline soils.

Saline-Alkali Soils

These soils have both soluble salts and exchangeable sodium. The soil reaction becomes strongly alkaline because of hydrolysis of exchangeable sodium. The physical condition of the soil is deteriorated. The EC of saturated extract is more than 4 dS m^{-1}, ESP is more than15 and pH is around 8.5. They are transitional soils that they may be converted into saline or into sodic soils.

Non-Saline Alkali Soils

Most of the Na^+ is in exchangeable form. EC of saturated extract is less than 4 dS m^{-1}. ESP is more than 15 and pH more than 8.5. Such soils are not in a good physical condition and the surface of these soils is sometimes black due to dispersion of organic matter and humus. Such soils if ploughed when wet, turn into slick furrow slice referred to as slick spots.

In addition to the parameters proposed by the USDA, Indian scientists considered the nature of soluble salts. Further the pH value of 8.5 is too high, as isoelectric pH for precipitation of $CaCO_3$ at which sodification starts is 8.2 and mostly the pH is associated with the ESP of 15 or more. The classification according to the Indian system is presented in Table 8.4.

Problems of Alkali Soils on Plant Growth

- Poor water and air permeability as a result of high dispersion of soil aggregates and clay particles
- Low availability of water due to poor conductance from the lower soil layers

- Seedling emergence and germination is greatly affected due to hard crust surface.
- Seed bed preparation is very difficult because it does not come to proper moisture conditions
- Deficiency of Ca and there is an antagonistic effect on K and Ca nutrition.
- Toxic concentration of HCO^{3-} and CO_3^{--} ions
- Decreased solubility of Fe and Zn.

Degraded Alkali Soils

The soil does not contain free $CaCO_3$. As a result of prolonged leaching under this condition, Na^- clay hydrolyses NaOH which combine with CO_2 or soil air and forms Na_2CO_3 (Alkaline condition). Theses soils formed through solodization.

Table 8.4: Indian System of Classification

Soil Characteristics	*Saline Soils*	*Alkali Soils*
pH	< 8.2	> 8.2
ESP	< 15	> 15
ECe	> 4 dS m^{-1}	Variable, mostly < 4 dS m^{-1}
Nature of soluble salts	Neutral, mostly Cl^-, SO_4^{2-}, HCO_3^- may be present but CO_3^{2-} is absent.	Capable for alkaline hydrolysis, preponderance of HCO_3^- and CO_3^{2-} of Na^+

Salinity Management in Agriculture

Different technologies are available for the reclamation and utilization of salt-affected soils, their strengths and weaknesses are briefly discussed under the following heads:

Saline and Sodic Soil Reclamation

The subsurface drainage technology, based on a network of concrete or PVC pipes installed at a particular spacing and depth below the soil surface, has proved highly successful in reclamation waterlogged saline soils in Haryana, Rajasthan, Gujarat, Punjab, Andhra Pradesh, Maharashtra, Madhya Pradesh and Karnataka states. The reclaimed soils put under crop production exhibit considerable increase in cropping intensity and yields. Substantial gains in terms of on-farm employment generation and productivity gains notwithstanding, the adoption of this technology at farmers' field is limited due to higher establishment costs and operational difficulties. The immense popularity of gypsum-based sodic soil reclamation package is evident from profitable cropping in once barren, alkali soils covering large area.

Bio-drainage of Waterlogged Lands

To overcome operational difficulties and social constraints in the adoption of sub-surface drainage for waterlogged saline lands, bio-drainage technology based on perennial trees has been developed (Ram *et al.*, 2011). Similar to energy-operated water pumps, bio-drainage involves the use of salt tolerant trees having high transpiration rates (*e.g.*, eucalyptus, popular, and bamboo) to arrest the processes of water logging and salt accumulation in canal commands (Singh, 2009). Available evidence indicates that appropriate land modification models could significantly enhance the potential of this technique.

Integrated Farming Model

As landholdings are becoming smaller with each passing day and higher resource use efficiencies are a pressing need for the sustainability of agriculture, an integrated model consisting of field and horticultural crops, fishery, cattle, poultry and beekeeping has been developed for 2 ha area for ensuring food and nutritional security of and regular incomes to a farm family consist of 4-5 members. This model substantially reduces the production costs by synergistic recycling of resources among different components (Singh, 2009). Similar models are in experimental stage of development for waterlogged sodic soils in Uttar Pradesh, highly saline black soils in Gujarat and coastal saline soils of West Bengal (Sharma and Chaudhari, 2012).

Microbial Consortia for Salt Stress Alleviation

To expedite the use of low cost crop growth bio-enhancers in stressful saline environments by sensitizing the farmers about their practical utility, a microbial bio-formulation named 'CSRBIO' has been developed. It acts as a soil conditioner and nutrient mobilizer and significantly increases the productivity of rice, banana, vegetables and gladiolus in sodic soils (Damodaran *et al.*, 2013)

Emerging Constraints and Future Perspectives

Sustainable cropping in salt-affected environments. Secondary salinity in irrigated commands of north-western India has caused huge land degradation (Datta and De Jong, 2002). Intensive rice-wheat rotation, faulty water management and clearing of vegetation seem to be the primary causes behind excessive salinity build-up in these regions. To prevent the problem of secondary salinity from becoming an ecological threat, an integrated approach based on precision land levelling, judicious resource use, crop diversification, adoption of salt tolerant trees and crops, organic nutrient management and conjunctive use of fresh and saline waters are recommended (Sharma *et al.*, 2011, Sharma and Chaudhari, 2012).

In near future, extreme climate variability could prove disastrous to cropping in saline environments as erratic rainfall, high temperature and droughts could increase the propensity of salt stress and could further undermine the productivity of these inherently vulnerable agro-ecosystems (Enfors and Gordon, 2007; Yeo, 1998).

In India, having 4.2 per cent global water resources, agriculture sector consumes the major chunk (~85 per cent) of the available fresh water. As adequate availability and good quality of water are essential for sustainable crop yields, anticipated water scarcity and increase in saline/sodic water resources would pose an additional burden in salt-affected regions (Singh, 2009). Future research agenda must focus on issues such as use of poor quality waters in soil reclamation, improvements in existing ground water recharge strategies, storage and subsequent use of rainwater through land modifications and refinements in existing technologies for optimizing saline/sodic water use in crop production (Sharma *et al.*, 2011; Sharma and Chaudhari, 2012). Application of novel technologies such as solute transport modelling and air-borne sensors would become necessary for speedy and precise mapping of salt affected soils. The time tested technologies of gypsum-based sodic soil reclamation and sub-surface drainage require substantial refinements for continued relevance under changing scenario. Future research experiments must strive to develop cheap and environment-friendly alternatives such as fly ash-, distillery spent wash- and press mud-based reclamation packages for sodic soils. The focus should now shift from breeding for salt tolerance to the development of multiple stress tolerant genotypes. The efficiency of resource conservation technologies must be demonstrated by multi-location trials. Research, extension and demonstration should be integrated into one fold for rapid dissemination of proven technologies to farmers' doorsteps. Targeted capacity building of farmers and other stakeholders would be equally important in technology-led sustainable management of salt-affected soil and water resources.

References

Arora Sanjay, Singh, Y.P. and Singh A. K. 2015. Training manual on Advanced Techniques for Bioremediation and Management of Salt Affected Soils. ICAR-CSSRI, RRS, Lucknow 226002 (UP).pp: 1-261.

Damodaran, T. *et al.*, 2013. Impact of social factors in adoption of CSR BIOA cost effective, ecofriendly biogrowth enhancer for sustainable crop production. *South Asian Journal of Experimental Biology*, 3: 158-165.

Datta, K. K. and De Jong, C. 2002. Adverse effect of waterlogging and soil salinity on crop and land productivity in northwest region of Haryana, India. *Agricultural Water Management*, 57: 223-238.

Enfors, E. I. and Gordon, L. J. 2007. Analysing resilience in dryland agro-ecosystems: a case study of the Makanya catchment in Tanzania over the past 50 years. *Land Degradation and Development*, 18: 680–696.

Ram, J. *et al.*, 2011. Biodrainage to combat waterlogging, increase farm productivity and sequester carbon in canal command areas of northwest India. *Current Science*, 100(11): 1673-1680.

Sharma, D. K. and Chaudhari, S. K. 2012. Agronomic research in salt affected soils of India: An overview. *Indian Journal of Agronomy*, 57: 175-185.

Sharma, D.K. *et al.*, 2011. CSSRI Vision 2030. Central Soil Salinity Research Institute, Karnal.

Singh, G. 2009. Salinity-related desertification and management strategies: Indian experience. *Land Degradation and Development*, 20: 367-385.

Wicke, B., *et al.*, 2011. The global technical and economic potential of bioenergy from salt- affected soils. *Energy and Environmental Science*, 4: 2669-2681.

Chapter 9

Economics of Resource Conservation Technologies for Enhancing Income and Maintaining Environment in Uttar Pradesh, India

O.P. Singh

Department of Agricultural Economics, Institute of Agricultural Sciences, Banaras Hindu University, Varanasi – 221 005, U.P.

1. Introduction

Food security for fast growing population and poverty alleviation on one hand and natural resource depletion, negative impact of climatic variability, mounting cost of inputs and volatile price of agricultural produce are the major challenges faced by the many Asian countries. Beside these challenges, some other factors faced by farmers are groundwater water depletion, reduction in total factor productivity, soil erosion, fast depletion of soil organic matter. The tillage of agricultural field as old as human civilisation and its first evidence was found dated back 3000 BC in Mesopotamia. After advent of mechanical power, farmers shifted from animal power to mechanical power for tillage operation and pumping irrigation water. The negative impact of intensive tillage are decline in soil organic matter, soil structural degradation, water and wind erosion, reduced water infiltration rates, surface sealing and crusting, soil compaction. Traditional agriculture, based on intensive tillage are responsible for soil erosion, surface and underground water pollution and more water consumption. The tragic dust storm in the Mid-western United

States in 1930 was wake-up call to how human interventions in soil management and ploughing led to unsustainable agriculture. In 1930, Edward H. Faulkner questioned the utility of intensive tillage (Faulkner, 1943). Since 1930, agricultural scientists were advocating farming community to reduce tillage. After development of seeding machine during 1940, it made sowing possible without soil tillage (Friedrich *et al.*, 2012).

Therefore, a paradigm shifts in farming practices through eliminating negative consequences of conventional agriculture (intensive ploughing, removal of all organic matters from soil and mono-cropping) is more crucial for future productivity gains. Conservation agriculture is a management system that maintains a soil cover through surface retention of crop residues with no till/zero and reduced tillage. Conservation agriculture is an efficient alternative to traditional agriculture provides excellent soil fertility, saves money, time and fossil-fuel. Resource Conservation Technologies (RCTs) are those that enhance farmers' productivity and profitability by making more efficient use of production resources and inputs. The key elements which characterize conservation agriculture are: [a] Minimum soil disturbance by adopting no-tillage and minimum traffic for agricultural operations; [b] Leave and manage the crop residues on the soil surface/crop filed; and [c] Adopt spatial and temporal crop rotations to derive maximum benefits from inputs and minimize adverse environmental impact. Looking the importance of RCTs, present study was an attempt to find out the economic and environmental benefit of RCTs in different climatic zones of Uttar Pradesh.

2. Research Methodology

2.1 Data Use and Sampling Procedure

Present study was based on primary and secondary data. Primary data was collected using pre-tested schedule through personal interview. The Uttar Pradesh was purposively selected for the present study. The Uttar Pradesh is divided into nine agro-climatic zones, out of these eight agro-climatic zones were selected for primary data collection. From each selected afro-climatic zone, one district was selected based on the adoption of resource conservation technologies. From each selected district, two blocks was selected based on highest and lowest area under RCTs. From each selected blocks two villages was selected based on highest and lowest adoption of RCTs. From each selected village 10 RCTs adopters and 10 non-adopters farmers were selected (Table 9.1).

2.2 Analytical Procedure

2.2.1 Cost of Cultivation

The cost of cultivation for wheat crop was estimated by using the costs concept suggested by the Commission on Agriculture Costs and Prices (CACP). Per hectare incremental cost of cultivation (Δ_{CC}) for wheat crop was estimated by using following method:

$$\Delta_{CC} = \vartheta_{RCT} - \vartheta_{C}$$

Table 9.1: Selection of Districts, Blocks, Villages and Sample Size

Name of Agro-climatic Zones	*Name of Districts*	*Name of Blocks*	*Name of Village*	*No. of Respondents*	
				Adopters	*Non-adopters*
1. Vindhyan	1. Mirzapur	1. Narayanpur	Bhurkura	10	10
		2. Jamalpur	Pirkhir	10	10
2. Central plain	2. Kaushambi	1. Newada	Kadirpur Newada	10	10
		2. Muratganj	Mohnapur	10	10
3. Western plain	3. Ghaziabad	1. Rajapur	Matiyala	10	10
		2. Bhojpur	Tahlata	10	10
4. North-eastern plain	4. Gorakhpur	1. Bhatahata	Pokharbhinda	10	10
		2.Brahmpur	Belwa	10	10
5. Eastern plain	5. Chandauli	1. Barhani	Barhani	10	10
		2. Sahebganj	Khilchi	10	10
6. Mid-western plain	6. Bareilly	1. Nawabganj	Vakaniya	10	10
		2. Vithari Chainpur	Sisaiya	10	10
7. Tarai and Bhabar	7. Bahraich	1. Risia	Patana Ghishiyari	10	10
		2. Shivapur	Itaha	10	10
8. South-western plain	8. Etah	1. Nidhauli Kalan	Gerhana	10	10
		2. Punehra	Awagarh	10	10
Total	**8**	**16**	**16**	**160**	**160**

Where Δ_{CC} is the incremental cost of cultivation (Rs. per hectare), ϑ_{RCT} is the total inputs cost used by the RCTs adopters (Rs. per hectare) and ϑ_{C} is total inputs cost used by the RCTs non-adopters (Rs. per hectare).

2.2.2 Economic Benefits

Economic benefits of RCTs were worked out using economic surplus model as suggested by Alston *et al.* (1995). The model is given below:

$$\Delta CS = P_0 Q_0 Z (1 + 0.5 Z_\eta)$$

$$\Delta PS = P_0 Q_0 (K - Z)(1 + 0.5 Z_\eta)$$

$$\Delta TS = P_0 Q_0 K (1 + 0.5 Z_\eta)$$

where,

P_0 = Base price of the commodity

Q_0 = Base Quantity

η = Absolute value of the price elasticity of demand

Z = K ε/(ε+ η); or the proportionate price reduction in the market where ε is the elasticity of supply

K = proportionate reduction in cost of production

ΔCS = Change in consumer surplus

ΔPS = Change in producer surplus

ΔTC = Change in total economic surplus

2.2.3 Carbon Emission

Environmental benefits realized by adoption of RCTs were quantified. Generally these benefits are reduction in carbon emission and improvement of carbon/organic content through residue management in soil. To find out the carbon emission following methodology was adopted:

1 liter Diesel = 2.6 kg of CO_2 (Jat *et al.*, 2006)

1 kg CO_2 = 0.27 kg of carbon (Paustian *et al.*, 2006)

2.2.4 Irrigation Water Use

For quantification of irrigation water used for wheat production following equation was used (Singh, 2004):

$$\theta(m^3) = I_n \times H_{pi} \times P_d$$

where, $\theta(m^3)$ s total irrigation water used for wheat production measured in m^3/ha, I_n is number of irrigation given to wheat crop during crop period; H_{pi} is the hours required to provide one irrigation, P_d is the pump discharge rate measured as m^3/hour.

2.2.5 Pump Discharge Rate

Farmers of the study area were using groundwater for irrigating wheat crop. For the quantification of irrigation water use following formula was used (Singh, 2004):

$$Pd_{(m^3/Hr)} = \frac{HP \times 75 \times Pe}{1000 \times DW} \times 3600$$

Where *Pd* is pump discharge rate measured in m^3 per hour; *HP* is the pump capacity; *Pe* is the pump efficiency; and *DW* is depth to water level plus head of delivery pipe measured in meter.

2.2.6 Physical Water Productivity

Physical water productivity (kg/m^3) for wheat crop was estimated using the data on crop yield and the estimated volume of irrigation water applied to wheat crop. The physical water productivity of wheat crop was estimated through following equation (Singh, 2004):

$$WP_{(kg/m^3)} = \frac{Q_{(kg)}}{\theta_{(m^3)}}$$

Where, *WP* is the physical water productivity of wheat crop measured as kg/m^3, $Q_{(kg)}$ is wheat yield measured in kg per hectare and $\theta(m^3)$ is the total irrigation water used for wheat crop during entire crop period measured in m^3/ha.

2.2.7 Net Economic Water Productivity

For estimation of net economic water productivity (Rs/m^3), data related to the crop yield (both main and by-product), farm harvest price and total input costs were used. The combined physical and economic water productivity for wheat production was estimated using following equation (Singh, 2004):

$$WP_{(Rs/m^3)} = \frac{NI_{(Rs)}}{\theta_{(m^3)}}$$

Where, $WP_{(Rs/m^3)}$ is the net combined physical and economic water productivity (Rs/m^3); $NI_{(Rs)}$ is the net income received from wheat crop (Rs/ha) and $\theta(m^3)$ is the total irrigation water used during entire period of wheat crop measured in m^3/ha.

3. Results and Discussion

3.1 Area under Conservation Agriculture

The area under conservation agriculture is presented in Figure 9.1. The total area under conservation agriculture was about 125 million hectare (Bhan and Behera, 2014). Out of total area under conservation agriculture, the highest area under conservation agriculture was in USA (21.2 per cent) followed by Brazil (20.4 per cent), Argentina (20.4 per cent), Australia (13.6 per cent), Canada (10.8 per cent), Russian Federation (3.6 per cent), China (2.5 per cent), Paraguay (1.9 per cent), Kazakhstan (1.3 per cent) and others (4.2 per cent). In India, nearly two million hectares are under conservation agriculture. The major conservation based technologies being adopted in India is zero-till in wheat crop in rice-wheat system of the Indo-Gangatic plain. In other crop, farmers are gradually shifting from intensive tillage to reduced/

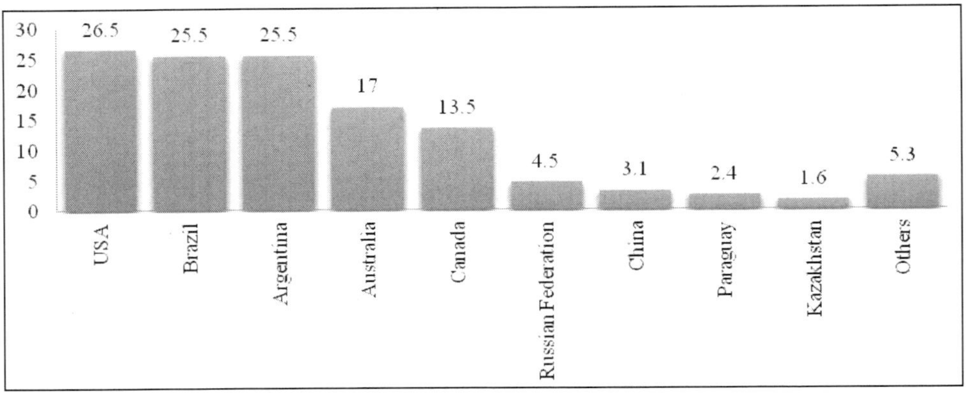

Figure 9.1: Area Under RTCs (million Ha).

zero-tillage tillage. The effort to develop and popularise the conservation agriculture have made through several State Agricultural Universities, ICAR institutions and Rice-Wheat Consortium for the Indo-Gangetic Plains. The spread of technologies is taking place in the irrigated regions of Indo-Gangetic plain where farmers are largely growing rice-wheat based cropping system.

3.2 Cost of Cultivation and Wheat Yield

In most of the agro-climatic zones of Uttar Pradesh, per hectare average cost of cultivation in case of RCTs adopters was lower as compared to non-adopters. In case of zero/reduce tillage, highest reduction in cost of cultivation was found in Vindhyan region (Rs. 7077) and lowest reduction was observed in North-eastern plain with Rs. 4151 as compared to non-adopters (Table 9.2). In Eastern plain, cost of cultivation for zero/reduce tillage adopters was higher as compared to non-adopters due to adoption of high yielding varieties of wheat seed. In case of rotavator, reduction in per hectare cost of cultivation was highest in Tarai and Bhabar agro-climatic zone with Rs. 11527, whereas it was lowest in South-western plain with Rs. 2309.04. In case of Mid-western plain, per hectare cost of cultivation was higher for rotavator adopter as compared to rotavator non-adopters.

After adoption of RCTs for wheat cultivation, farmers were getting higher wheat yield as compared to RCTs non-adopters. In case of zero/reduce tillage, per hectare highest incremental wheat yield over conventional method was observed for Western plain (6.21 qts.) and lowest for Vindhyan agro-climatic zone with 2.61 quintal. In case of rotavator, per hectare highest incremental wheat yield was observed for Mid-western plain (8.6 qts) and lowest in case of Tarai and Bhabar agro-climatic zone with 7.12 quintal over conventional method.

Per hectare higher wheat yield obtained by RCTs adopters was due to incorporation of crop residues into agricultural field, which has positive and significant impact on soil physical, chemical and biological properties. Mixing of crop residue in agricultural field enhances soil organic carbon content, increases

Table 9.2: Cost of Cultivation and Yield of Wheat Crop under different RCTs

Name of the Agro-Climatic Zones	Cost of Cultivation (Rs/Ha)		Difference in Cost of Cultivation (Rs/Ha)	Crop Yield (Qts/Ha)		Incremental Yield Cenefit (Qts/Ha)
	RCTs Adopter	RCTs non-adopters		RCTs Adopter	RCTs non-adopters	
	Zero/Reduce Tillage					
1. Vindhyan	51361.65	58438.73	7077.08	35.59	32.98	2.61
2. Central plain	39213.50	45744.02	6530.52	35.45	29.50	5.95
3. Western plain	51315.21	57269.99	5954.78	49.88	43.67	6.21
4. North-eastern plain	48459.35	52609.86	4150.51	35.75	30.93	4.82
5. Eastern plain	48695.03	47320.56	-1374.47	32.01	27.16	4.85
	Rotavator Tillage					
6. Mid-western plain	39835.19	39027.52	-807.67	44.08	35.48	8.60
7. Tarai and Bhabar	45666.83	57194.71	11527.88	44.04	36.92	7.12
8. South-western plain	48581.62	50890.66	2309.04	45.36	37.21	8.15

N, P, K availability, prevents nitrogen leaching in deeper soil, augments biodiversity in the soil, moderates soil temperature, reduces evaporation, improves biological activity and provides more conducive environment for root development, which leads to augmentation of wheat yield (Singh and Sharma, 2005; Sidhu and Beri, 2005; Misra *et al.*, 1996; Singh and Meena, 2013).

3.3 Gross and Net Income from Wheat

The RCTs reduce the use of human and mechanical labour. It advances in sowing time, helps in reduction in fossil fuel and farm power use for different agricultural operations which leads to reduction in cost of cultivation as compared to conventional method (Singh and Sharma, 2005; Laxmi and Mishra, 2007). The reduction in cost of cultivation varies with agro-climatic regions, technology adopted by farmers and crops grown under RCTs (Sarwar and Goheer, 2007; Sikka *et al.*, 2005; Grover *et al.*, 2005).

After adoption of RCTs in the study area, sample farmers were receiving higher gross and net income as compared to RCTs non-adopters. It is due to the lower use of inputs in wheat cultivation and higher wheat production (both main and by-products). In case of zero/reduce tillage adopters, per hectare highest incremental gross income was observed in Western plain (Rs. 11118) and lowest in Vindhyan agro-climatic zone with Rs. 4692 over conventional method of wheat cultivation. In case of rotavator, highest per hectare gross incremental benefit was observed in South-western plain (Rs. 17139.91) and lowest in Tarai and Bhabar agro-climatic zone (Rs. 13244.81) over conventional method of wheat cultivation (Table 9.3).

Per hectare net incremental income received by zero/reduce tillage adopters was highest in Western plain (Rs. 17073) and lowest in Eastern plain (Rs. 7590.23) agro-climatic zones over zero-tillage non-adopters. In case of rotavator adopters per hectare highest net incremental benefit was received in Tarai and Bhabar (Rs. 24773) and lowest in Mid-western plain agro-climatic zone with Rs. 13276.60 over rotavator non-adopters (Table 9.3).

3.4 Economic Benefits of RCTs Adoption

The economic benefits of RCTs are reduction in cost of human labour, cost of machine labour, cost of seed, cost of fertiliser, cost of pesticide, cost of irrigation, cost of harvesting, increase in wheat yield (both main and by-products) and saving of diesel cost. Per hectare economic benefits due to adoption of zero/reduce tillage in the study area were estimated and it was found to be highest in Western plain (Rs. 20537.43), whereas it was lowest in Eastern plain with Rs. 10426.91 (Table 9.4). In case of rotavator adopters, per hectare economic benefit was found to be highest in Tarai and Bhabar agro-climatic zone with Rs. 26130.17, whereas it was lowest in Mid-western plain with Rs. 18701.06 (Table 9.4).

3.5 Reduction in Carbon Emission

Many past researchers reported that after adoption of RCTs, the air pollution reduces and mitigate environmental degradation by management of crop residue and carbon sequestration (Akhtar, 2006; Laxmi and Mishra, 2007). RCTs reduce

Table 9.3: Gross and Net Income from Wheat Crop under Different RCTs

Name of Agro-climatic Zones	*Gross Income (Rs/Ha)*		*Incremental Gross Income (Rs/Ha)*	*Net Income (Rs/Ha)*		*Incremental Net Income (Rs/Ha)*
	RCTs Adopters	*RCTs non-adopters*		*RCTs Adopters*	*RCTs Non-adopters*	
	Zero/Reduce Tillage					
1. Vindhyan	66153.55	61461.97	4691.58	14791.90	3023.24	11768.66
2. Central plain	56051.10	48168.97	7882.13	16837.60	2424.94	14412.66
3. Western plain	89335.65	78217.56	11118.09	38020.44	20947.57	17072.87
4. North-eastern plain	60948.79	53709.01	7239.78	12489.44	1099.15	11390.29
5. Eastern plain	57984.57	49019.87	8964.70	9289.53	1699.30	7590.23
	Rotavator Tillage					
6. Mid-western plain	71684.99	57600.71	14084.28	31849.79	18573.19	13276.60
7. Tarai and Bhabar	81855.99	68611.18	13244.81	36189.16	11416.47	24772.69
8. South-western plain	94300.95	77139.91	17139.91	45719.32	26249.25	19470.07

Table 9.4: Economic Benefit of RCTs (Rs/Ha)

Economic Benefits Due to Reduction in	*Name of the Agro-climatic Zones of Uttar Pradesh*							
	Vindhyan	*Central Plain*	*Western Plain*	*North-eastern Plain*	*Eastern Plain*	*Mid-western Plain*	*Tarai and Bhabar*	*South-western Plain*
	Zero/Reduce Tillage						*Rotavator Tillage*	
Cost of human labour	1363.65	783.06	1378.94	131.20	548.58	355.86	50.78	-32.56
Cost of machine labour	3042.88	3199.42	3408.12	2504.74	1458.11	2954.00	2931.56	1461.95
Cost of seed	893.73	681.96	674.07	645.05	-309.69	629.54	109.27	82.14
Cost of fertilizer	556.24	108.05	411.61	425.64	-1507.42	49.35	1391.72	-384.19
Cost of pesticide	16.75	2.50	27.50	-12.50	25.00	0.00	3.75	0.00
Cost of irrigation	568.98	848.57	1615.37	663.48	593.02	1195.12	2332.22	1840.23
Cost of harvesting	129.87	112.53	29.26	-479.44	10.90	-2191.78	3167.68	68.11
Due to yield benefits	4691.59	7882.13	11118.09	7239.78	8964.70	14084.27	13244.81	17161.04
Due to diesel saving*	1748.41	1564.16	1874.47	1422.91	643.72	1624.70	2898.37	536.05
Total	13012.10	15182.37	20537.43	12580.85	10426.91	18701.06	26130.17	**20732.76**

*: Cost of diesel is Rs. 55 per litre.

diesel consumption for land preparation and sowing of wheat crop which leads to reduction in carbon emission (Pal *et al.*, 2010). One litre diesel burning generates 2.6 kg of CO_2 (Jat *et al.*, 2006) and one kg CO_2 is equal to 0.27 kg of carbon (Paustian *et al.*, 2006). It is clearly noticed from Table 9.5 that after adoption of RCTs, diesel consumption for land preparation, sowing of wheat crop and groundwater pumping using diesel operated pumps substantially reduced, which leads to reduction in the carbon emission.

Table 9.5: Carbon Emission from Wheat Crop per Hectare

Name of the Agro-climatic Zones	*RCTs Adopters*		*RCTs Non-adopters*		*Reduction in*	
	Diesel Use (Lts)	*Carbon Emission (kgs)*	*Diesel Use (Lts)*	*Carbon Emission (kgs)*	*Diesel Use (Lts)*	*Carbon Emission (kgs)*
	Zero/Reduce Tillage					
1. Vindhyan	64.87	45.55	96.67	67.86	31.80	22.31
2. Central plain	8.25	6.21	37.29	26.18	29.04	19.97
3. Western plain	25.07	17.60	59.15	41.52	34.08	23.92
4. North-eastern plain	57.78	52.41	83.09	71.08	25.31	18.67
5. Eastern plain	15.00	10.53	26.70	18.75	11.70	8.22
	Rotavator Tillage					
6. Mid-western plain	103.39	72.58	146.47	102.83	43.08	30.25
7. Tarai and Bhabar	85.50	60.02	138.68	97.35	53.18	37.33
8. South-western plain	105.01	73.72	127.22	89.30	22.21	15.58

In case of zero/reduce tillage, reduction in diesel consumption per hectare was highest in Western plain (34.08 lts) and lowest in Eastern plain with 11.70 lts as compared to non-adopters. After adoption of zero/reduce tillage, per hectare carbon emission reduced from wheat cultivation in Western plain with 23.92 kg which was highest and lowest in Eastern plain region with 8.22 kg as compared to zero/reduce tillage non-adopters.

In case of rotavator, per hectare reduction in diesel consumption was highest in Tarai and Bhabar (53.18 lts) and lowest in South-western plain as compared to non-adopters (Table 9.5). Per hectare reduction in carbon emission was maximum in Tarai and Bhabar agro-climatic zone with 37.33 kg and lowest in South-western region with 15.58 kg as compared to rotavator non-adopters. The difference in mean value of carbon emission was found to be significant in all the agro-climatic zones.

3.6 Irrigation Water Use

Past researchers reported that, RCTs have enabled farmers to sow their wheat crop just after paddy harvesting without any pre-sowing irrigation and it will further reduce the irrigation water in subsequent irrigation (Hobbs *et al.*, 2000; Malik *et al.*, 2005; Akhtar, 2006; Pal *et al.*, 2010). Irrigation water used during entire period of wheat cultivation by sample farmers in the study area suggests that RCTs adopters were using less irrigation water as compared to RCTs non-adopter in different agro-climatic zones of Uttar Pradesh (Figure 9.1).

In case of zero/reduce tillage adopters, per hectare irrigation water used was highest in Western plain (2235.44 m^3) and lowest in Vindhyan agro-climatic zone with 700.57 m^3, whereas in case of non-adopters it was highest in Western plain (3582.95) and lowest in Vindhyan agro-climatic zone with and 772.26 m^3 per hectare respectively (Figure 9.1). After adoption of zero/reduce tillage for wheat cultivation, per hectare irrigation water saved was highest in Western plain with a tune of 1347.51 m^3 and lowest in Vindhyan zone with 71.69 m^3. In case of rotavator adopters, per hectare irrigation water used was highest for wheat cultivation in the study area which was estimated to be 4336.17 m^3 and lowest (2628.75 m^3) in Tarai and Bhabar and Mid-western plain respectively, whereas in case of non-adopters, it was highest 5707.25 m^3 and lowest (3218.91 m^3) for Tarai and Bhabar and Mid-western plain respectively (Figure 9.2). After adoption of rotavator, per hectare highest irrigation water saved was highest in Tarai and Bhabar agro-climatic zone with 1371.08 m^3 and lowest in Mid-western plain with 590.16 m^3.

3.7 Agronomic and Net Economic Water Productivity

The agronomic water productivity (kg/m^3) and net economic water productivity (Rs/m^3) was estimated for RCTs adopter and non-adopters for different agro-climatic zones of Uttar Pradesh. In case of zero/reduce tillage adopters, highest agronomic water productivity was observed in Vindhyan agro-climatic zone (5.08 kg/m^3) and lowest in Eastern plain (1.84 kg/m^3), whereas in case of non-adopters it was highest in Vindhyan agro-climatic zone (4.27 kg/m^3) and lowest in Western plain (1.22 kg/m^3). In case of rotavator adopters, highest agronomic water productivity was observed for Mid-western plain and lowest in Tarai and Bhabar agro-climatic zone with 1.68 and 1.02 kg/m^3 respectively, whereas in case of non-adopters it was highest in Mid-western plain (1.10 kg/m^3) and lowest in Tarai and Bhabar agro-climatic zone with 0.65 kg/m^3 (Table 9.6).

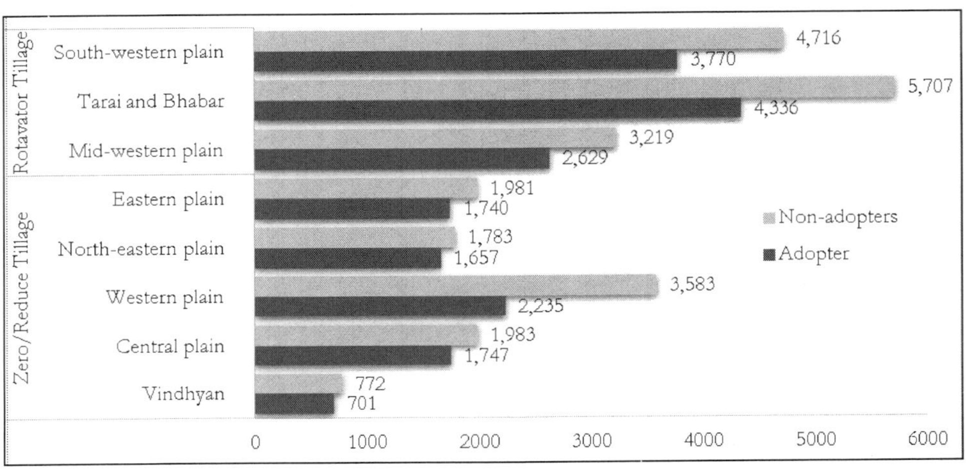

Figure 9.2: Irrigation Water Use for Wheat Crop (m^3/ha).

Table 9.6: Agronomic and Net Economic Water Productivity

Name of the Agro-climatic Zones	Agronomic Water Productivity (kg/m³)		Incremental Agronomic Water Productivity (kg/m³)	Net Economic Water Productivity (Rs/m³)		Incremental Net Economic Water Oroductivity (Rs/m³)
	RCTs Adopters	RCTs Non-adopters		RCTs Adopters	RCTs Non-adopters	
	Zero/Reduce Tillage					
1. Vindhyan	5.08	4.27	0.81	21.11	3.91	17.2
2. Central plain	2.03	1.49	0.54	9.64	1.22	8.42
3. Western plain	2.23	1.22	1.01	17.01	5.85	11.16
4. North-eastern plain	2.16	1.73	0.43	7.54	0.62	6.92
5. Eastern plain	1.84	1.37	0.47	7.88	3.03	4.85
	Rotavator Tillage					
6. Mid-western plain	1.68	1.10	0.58	12.12	5.77	6.35
7. Tarai and Bhabar	1.02	0.65	0.37	8.35	2.00	6.35
8. South-western plain	1.20	0.79	0.41	12.13	5.57	6.56

Net economic water productivity (Rs/m^3) in wheat crop for different agro-climatic zones for RCTs adopters and non-adopters were estimated. In case of zero-tillage adopters, the highest net economic water productivity was observed in Vindhyan agro-climatic zone (Rs. 21.11/m^3) and lowest in North-eastern plain climatic zone (Rs. 7.54/m^3). In case of non-adopters, highest net economic water productivity was found in Western-plain and lowest in North-eastern plain agro-climatic zone with Rs. 5.85 and Rs. 0.62 per m^3 respectively. In case of rotavator adopters, the highest net economic water productivity was found in South-western plain and lowest in Tarai and Bhabar agro-climatic zone. In case of non-adopters, the highest net economic water productivity was observed in Mid-western plain and lowest in Tarai and Bhabar agro-climatic zone (Table 9.6).

4. Summary and Conclusions

Among the different options of RCTs, farmers in different agro-climatic zones of Uttar Pradesh are adopting and using two technologies *i.e.* zero/reduce tillage and rotavator for land preparation and sowing of wheat crop. On an average cost of cultivation of wheat crop was lower for zero/reduce tillage and rotavator adopters as compared to non-adopters. Per hectare wheat yield was higher for zero-tillage and rotavator adopters as compared to non-adopters. Per hectare gross and net return from the wheat under zero-tillage and rotavator was higher than the non-adopters. It was due to higher crop yield and reduction in inputs used in the study area. The economic benefits due to adoption of RCTs were higher in all the agro-climatic zones of Uttar Pradesh. After adoption of RCTs, diesel consumption and carbon emission reduced substantially in the study area. Per hectare irrigation water used for wheat cultivation was lower in case of RCTs adopters as compared to non-adopters. The agronomic and net economic water productivity was higher for RCTs adopters as compared to non-adopters. The major constraints associated with the adoption of RCTs were high cost of machine, non-availability of machine on time and dearer custom haring. For the adoption of RCTs on large scale, government should provide more subsidy for the purchase of zero-tillage and rotavator machine.

References

Akhtar, M. R. (2006). Impact of resource conservation technologies for sustainability of irrigated agriculture in Punjab-Pakistan, *Journal of Agricultural Research (Lahore)*. 44(3): 239-254.

Alston, J.M; Norton, G.W. and Pardey, P.G. (1995). *Science under scarcity: Principles and practice for agricultural research evaluation and priority setting*, Cornell University Press, Ithaca.

Bhan, Suraj and U. K. Behera (2014) Conservation agriculture in India – Problems, prospects and policy issues, *International Soil and Water Conservation Research*, 2(4): 1-12.

Faulkner, E. H. (1943). *Plowman's folly*, Michael Joseph Ltd., 26 Bloomsbury Street, W.C.1, London.

Friedrich, T., Derpsch, R. and Kassam, A. H. (1912). Global overview of the spread of conservation agriculture, *Field Action Science Reports,* 6: 1-7.

Grover, D.K., Joginder Singh, Ranjeet Singh and S.S. Dhillon (2005). Socio-economic impact assessment of bed planting technology in Punjab, In Abrol, I.P., R.K. Gupta and R.K. Malik (ed.) *Conservation Agriculture: Status and Prospects,* Centre for Advancement of Sustainable Agriculture, National Agricultural Science Center (NASC) Complex, New Delhi: 155-162.

Hobbs, P. R. and Gupta, R. K. (2004). Problems and challenges of no tillage for the rice-wheat system of the Indo-Gangatic plains in south Asia. In Lal, R., Hobbs, P., Uphoff, N. and Hasen D. O. (eds) *Sustainable agriculture and the rice-wheat system,* Ohio State University/Marce Dekker, Columbus, 101-119.

Jat, M.L., Sharma, S.K. and Singh, K.K. (2006). Conservation Agriculture for Sustainable Farming in India. Paper Presented in Winter School Training at Department of Agronomy, Tamil Nadu Agricultural University, Coimbatore, January 21.

Laxmi, V. and Mishra, V. (2007). Factors affecting the adoption of resource conservation technology: Case of zero tillage in rice-wheat farming systems, *Indian Journal of Agricultural Economics,* 62(1): 126-138.

Malik, R. K., Yadav, Ashok and Singh, Sher (2005). Resource conservation technologies in rice-wheat cropping system of Indo-Gangetic plains, In Abrol, I.P., R.K. Gupta and R.K. Malik (ed.) *Conservation Agriculture: Status and Prospects,* Centre for Advancement of Sustainable Agriculture, National Agricultural Science Centre (NASC) Complex, New Delhi: 13-22.

Misra, R.D., Pandey, D.S. and Gupta, V.K. (1996). Crop residue management for increasing the productivity and sustainability in rice-wheat system, In: Abstract of poster sessions. 2nd International Crop Science Congress, pp. 42, National Academy of Agricultural Sciences and ICAR, New Delhi, India.

Pal, S., Sekar, I. and Kar, A. (2010). Environmental impact of resource conservation technology: the case of zero-tillage in the rice -wheat system of south Asia, report to the CGIAR standing panel on impact assessment, Division of Agricultural Economics, Indian Agricultural Research Institute, New Delhi: 1-23.

Paustian, K., Antle, M., Sheehan, J. and Eldor, P. (2006). *Agriculture's role in greenhouse gas mitigation,* Washington, DC: Pew Center on Global Climate Change.

Sarwar, M.N., and Goheer, M.A. (2007). Adoption and impact of zero-tillage technology for wheat in rice-wheat system: water and cost saving technology: A case study from Pakistan (Punjab). Paper presented at the International Forum on Water Environmental Governance in Asia, March 14-15. Bangkok, Thailand.

Sidhu, B.S. and Beri, V. (2005). Experience with managing rice residues in intensive rice-wheat cropping system in Punjab, In Abrol, I.P., R.K. Gupta and R.K. Malik (ed.) *Conservation Agriculture: Status and Prospects,* Centre for Advancement of Sustainable Agriculture, National Agricultural Science Center (NASC) Complex, New Delhi: 55-63.

Sikka, A.K., Khan, A.R., Singh, S.S. and Subash, N. (2005). water management imperatives in zero/reduced till, In Abrol, I.P., R.K. Gupta and R.K. Malik (ed.) *Conservation Agriculture: Status and Prospects*, Centre for Advancement of Sustainable Agriculture, National Agricultural Science Center (NASC) Complex, New Delhi: 93-98.

Singh, K.K. and Sharma, S.K. (2005). Conservation tillage and crop residue management in rice-wheat cropping system, In Abrol, I.P., R.K. Gupta and R.K. Malik (ed.) *Conservation Agriculture: Status and Prospects*, Centre for Advancement of Sustainable Agriculture, National Agricultural Science Center (NASC) Complex, New Delhi: 23-32.

Singh, K.M. and Meena, M.S. (2013). *Economics of conservation agriculture: An overview*, Munich Personal RePEc Archive: 1-18.

Singh, O. P. (2004). Water productivity of milk production in North Gujarat, Western India, *Proceedings of the 2nd Asia Pacific Association of Hydrology and Water Resources (APHW) Conference*, Vol. 1: 442-449.

Chapter 10

Processing and Value Addition of Fruits and Vegetables for Doubling Farmer's Income

Pawan Singh Gurjar

Scientist (Fruit Science), Division of Post Harvest Management, ICAR-Central Institute for Subtropical Horticulture, Rehmankhera, Lucknow

Fruits and vegetables account for nearly 90 per cent of the total horticulture production in the country. India is now the second largest producer of fruits and vegetables in the world and is the leader in several horticultural crops, namely mango, banana, papaya, cashew nut, areca nut, potato, and okra. Its horticulture production has increased significantly over the last two decades and as per the final estimates, by 2014-15, it has increased to about three times (2.87) since 1991-92 and to about twice (1.90) compared to the production in 2001-02. The pace with production has increased, processing and value addition infrastructure has not been developed. At present, only 4.6 per cent of total fruits and vegetable production is processed (APEDA, 2014). Being perishable in nature, fruits and vegetable should be either processed or immediately marketed after harvesting. Lack of processing facility force farmers to sold their produce at throw away prices which result in low income to the farmers and high post harvest losses. Processing and value addition of fruits and vegetables can play a major role in doubling farmers income and developing entrepreneurship in rural areas.

Fruits and vegetables can be preserved by using various preservation techniques according to their feasibility and cost effectiveness. Different preservation techniques are listed below:

1. Removal of Moisture or Drying

Probably the oldest method of food preservation practiced by mankind. The removal of moisture prevents the growth and reproduction of micro-organisms causing decay as well as minimizes many of the moisture mediated biochemical deterioration reactions in the food. A number of drying methods such as sun drying, solar drying, cabinet drying and oven drying can be employed for drying of fruits and vegetables. Osmotic dehydration is another process of water removal by immersion of water containing cellular solids (fruits/vegetables) in a concentrated aqueous solution of sugar/salt. This results in intermediate moisture product with lower water activity. At low water activity, most of the chemical reactions which deteriorate the food, the growth and toxins production by microorganisms are ceased. Osmotic dehydration process has been applied to many fruits and vegetables *viz.* apple, apricot, banana, carrot cherry, citrus fruits, grapes, guava, papaya, mango, potato, *etc.*

Spray Drying

It is a method of producing a dry powder from a liquid or slurry by rapidly drying with a hot gas. This is the preferred method of drying of many thermally-sensitive materials such as foods and pharmaceuticals. A spray dryer takes a liquid stream and separates the solute or suspension as a solid and the solvent into a vapor (Figure 10.1). The liquid input stream is sprayed through a nozzle into a hot vapor stream and vaporized. Spray drying often is used as an encapsulation technique by the food and other industries. For example instant mix drinks. Spray drying has many applications in processing industry for making fruits and vegetable powders, juice powders, milk powder, coffee, tea, egg powder, vitamins, enzymes, neutraceuticals *etc.*

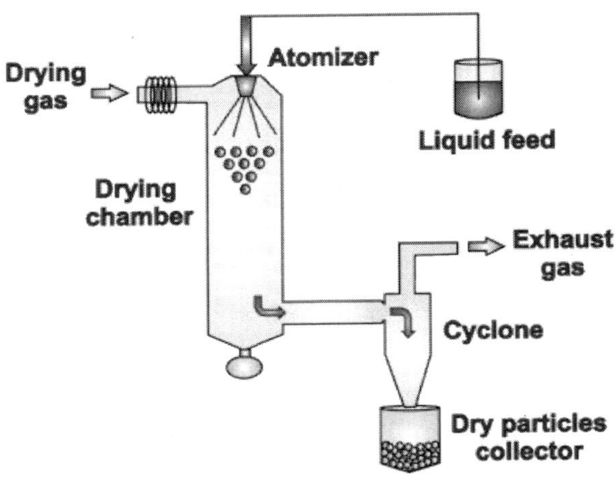

Figure 10.1: Spray Drying Process.

Freeze Drying

This process also known as lycophylization or cryodessication, is a low temperature dehydration process which involves freezing the product, lowering pressure, then removing ice by sublimation. Freeze drying results in high quality products because the low temperature used in the processing protect from loss in quality and nutrients. The original shape of the product is maintained and the quality of the rehydrated product is excellent.

Vacuum Drying

Vacuum drying is a process in which materials are dried in a reduced pressure environment, which lowers the heat needed for rapid drying. Vacuum dryers offer low- temperature drying of thermo labile materials and are suitable for solvent recovery from solid products containing solvents. Heat is usually supplied by passing steam or hot water through hollow shelves (Figure 10.2). Drying temperatures can be controlled for the major part of the drying cycle, the material remains at the boiling point of the wetting agent. The pressure maintained in vacuum drying is generally 0.03–0.06 atm and the boiling point of water is 25-30°C. Drying times are long, usually about 12 to 48 h.

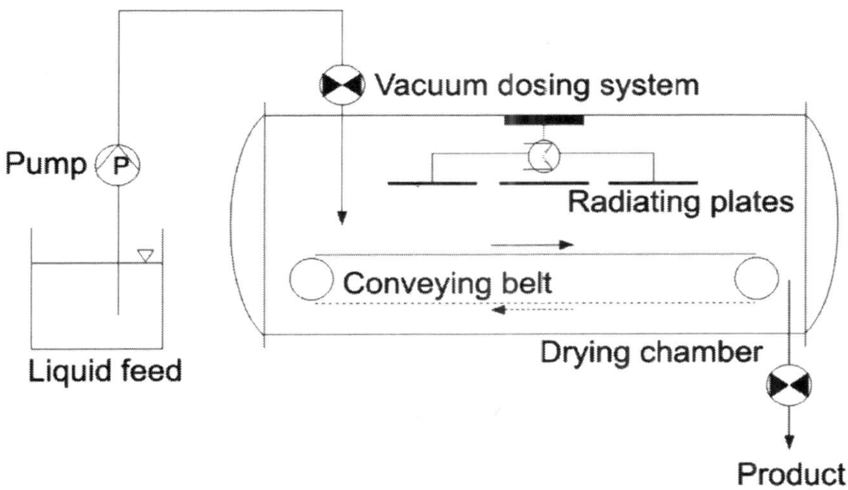

Figure 10.2: Vacuum Drying Process.

2. Thermal Processing

Heat is widely used in preservation of fruits and vegetables. Both microbial and enzymatic spoilage in produce can be checked. Canning of fruits and vegetables is based on thermal processing where complete destruction of micro-organisms achived by cooking at high temperature. Almost all fruits being acidic in nature processed in sugar syrup at 100 °C (in boiling water) while vegetables except tomato generally non-acidic in nature processed at high teperature-120-121 °C and high pressure. Pasteurization is another process in which fruit and vegetable juices sterilized by heating at 85-90 °C for 30 minutes to kill spore forming bacteria, mould spore and enzymes inactivation. Flesh pasteurization at a high temperature for short time only (90-95 °C for 1 minute). Blanching is the process of heating in hot water at 88° C (190° F) for two to five minutes or with steam in a conveyor at 100° C (212° F) for one-half to one minute. Blanching inactivates natural enzymes that would cause discoloration and off-flavours and aromas in vegetables.

3. Preservation with Natural Preservatives

Sugar, salt and oil are considered natural preservatives. The principle of this technology is to add sugar in a quantity that is necessary to augment the osmotic pressure of the product's liquid phase at a level which will prevent microorganism development. In concentrations of 60 per cent in the finished products, the sugar generally assures food preservation. Jam, jelly, marmalade, preserved and candid fruits are preserved by adding sugar. Minimum percentage of the fruit in the final product should be 45 per cent. Jam is prepared by boiling the fruit with a sufficient quantity of sugar (minimum 68 per cent) to a reasonably thick consistency. Jelly is prepared by fruit juice or clear water extract of fruit. Preserve or *murabba* is made by cooking or heat processing of raw or mature, peeled or punctured and cut fruits and vegetables in sugar syrup. Preserved when drained free from syrup and dried is called candy which is prepared largely from karonda, aonla, petha, cherry *etc.* Candies coated with thin transparent coating of sugar crystal are known as crystallized fruits.

4. Preservation by Fermentation

Fermentation is a low cost technology for preservation of fruits and vegetables. Ethanol fermentation is the process in which glucose is converted to ethanol (C_2H_5OH) and carbon dioxide (CO_2). Yeasts perform this conversion in the absence of oxygen therefore alcoholic fermentation is considered as an anaerobic process. Ethanol fermentation is used for making fruit cider and wines. Lactic acid (LA) fermentation of vegetables and fruits is a common practice to maintain and improve the nutritional and sensory features of food commodities. The basic mechanism of the preservation of foods through lactic acid fermentation is the production of acid, chiefly by lactic acid bacteria, which lowers the pH to a level at which most of the spoilage-causing microorganisms, cannot grow, and thus, the food is preserved. Other than this, food substrates overgrown with desirable, edible microorganisms become resistant to invasion by spoilage, toxic or food poisoning microorganisms. Vegetable pickles, sauerkraut (cabbage product), Kanji from black carrot juice and probiotic drinks is being manufactured by harnessing the principle of lactic acid bacteria fermentation

5. High Pressure Processing (HPP)

High pressure processing or high hydrostatic processing is a method of preserving and sterilizing food, in which a product is processed under high pressure leading to the inactivation of certain enzymes and microorganism in the food. HPP has limited effects on covalent bonds within the food products, thus maintaining both the sensory and nutritional aspects. During HPP, more than 50000 pounds per square inch applied for around 15 minutes leading to inactivation of yeast, mould and bacteria. Under HPP food products are sealed and placed into a steel compartment containing water and pumps are used to create pressure. This technology works especially well in acidic foods such as fruits and yoghurt because pressure tolerant spores are not able to live in environment with low pH levels. Self-life extended while keeping taste and nutritional properties of fresh fruits and vegetables through HPP. There are negligible chances of cross contamination because it applied post packaging. Product developed from HPP is rich in vitamins, anti-oxidants and thermo sensitive anti mutagenic components. Provide opportunities for new product development.

6. Preservation by Low Temperature

Individual Quick Freezing (IQF) is a freezing method used in preservation of fruits and vegetables. The products commonly frozen with IQF technologies are typically smaller pieces of food products and can range from all types of berries, fruits and vegetables diced or sliced. One of the main advantages of this method of preparing frozen food is that the freezing process takes only a few minutes. The short freezing prevents formation of large ice crystals in the product's cells, which destroys the membrane structures at the molecular level. This makes the product keep its shape, colour, smell and taste after defrost. There are 2 main IQF technologies: mechanical IQF freezers and cryogenic IQF freezers. Mechanical IQF freezers work on the principle of cold air circulation, which flows from underneath the bed plate or transportation belt with the help of fans. The cold airflow keeps passing through the pieces of product in circular motions while the product is also advancing through the freezer towards the exit end. Cryogenic IQF freezers immerse the product in liquid nitrogen at very low temperatures, freezing it very rapidly while continuously moving the product to avoid block or lump formation.

7. Preservation by Irradiation

This technique involves treatment of fruits and vegetables by subjecting them to ionizing radiation. Cobalt-60, Cesium-137 metals and machine generated beta or x-rays are used as source of radiation. Irradiation is used to reduce or eliminate the risk of food-borne illnesses, prevent or slow down spoilage, arrest maturation or sprouting and as a treatment against pests. Depending on the dose, some or all of the pathogenic organisms, present are destroyed, slowed down, or rendered incapable of reproduction (Table 10.1).

Table 10.1: Irradiation Dose for different Purposes

Sl.No.	*Applications*	*Dose (kGy)*
1	Inhibit sprouting (potatoes, onions, yams, garlic)	0.06 - 0.2
2	Delay in ripening (strawberries, potatoes)	0.5 - 1.0
3	Prevent insect infestation (grains, cereals, coffee beans, spices, dried nuts, dried fruits, dried fish, mangoes, papayas)	0.15 - 1.0
4	Extend shelf-life (raw and fresh fish, seafood, fresh produce, refrigerated and frozen meat products)	1.0 - 7.0
5	Reduce risk of pathogenic and spoilage microbes (meat, seafood, spices, and poultry)	1.0 - 7.0
6	Sterilization of packaging material	10-25

How to Establish Small Scale Processing Industry and Food Safety Laws

Small scale processing unit can be established in rural areas keeping into consideration the availability of raw material, available resources and marketing facility. New small scale processing unit may be set up in rural areas by young entrepreneurs or self help groups based on the following preservation techniques/products.

- Minimal processing of fruits and vegetables
- Processing and value addition of raw mango
- Processing of ripe mango and guava
- Aonla products
- Spices powder (Turmeric, chilli, coriender)
- Sun/solar dried vegetables
- Fruit and vegetable powders
- IQF of fruits and vegetables

Requirements for Establishment of Processing Unit

One should consider availability of raw material for processing, financial requirements, uninterrupted water and electricity supply and marketing of processed products for establishment and successful running of processing unit. FSSAI license issued by food safety department of concern state government is necessary for building consumers' confidence in product and it will also facilitates marketing.

Food Safety Laws

The Food Safety and Standards Authority of India (FSSAI) was established to implement the rules and regulations of FSS act under the aegis of Ministry of Health and Family Welfare, Government of India which is headquartered in New Delhi. It ensures the availability of safe food products for human consumption by laying down a set of safety standards in lieu to regulate the manufacturing, distributing and storing of the food products. It is thus responsible for promoting the public health among the masses by keeping a regular check.

To Whom it Required

FSSAI Registration or Food License is required for all type of the food related business. Even if you are thinking to starting a food related business by way of food processing or manufacturing or distributor then its required from starting irrespective of the turnover.

Types of FSSAI Food Licenses

FSSAI issues three different types of food licenses

Figure 10.3: Requirements for Establishment of Fruits and Vegetable Processing Unit.

Basic FSSAI Registration

The petty and small-sized food business operators like manufacturers, transporters, storage units, distributer, marketers, retailers, *etc.* are required to acquire a Basic FSSAI Registration which is issued by the State Government for a minimum period of one year and a maximum of 5 years. It is mostly for the units having an annual turnover of less than 12 lakh. The rest depends on the eligibility; the FBO can thus fall under either the State FSSAI License or the Basic Registration.

State FSSAI License

The authority has guided the food business operators, having an annual turnover of more than 12 lakh, like small to medium-sized manufacturers, storage units, transporters, marketers, retailers, distributors, *etc.* to obtain the State FSSAI License. It is issued by the respective State Government having a minimum validity of one year and the maximum of 5 yrs.

Central FSSAI License

Food business operators who have a turnover greater than 20 crores like large manufacturers, 100 per cent export oriented units, importers, operators in the Central

Government agencies, airports, seaports, *etc.* are enforced to obtain a Central Food License issued by the Central Government. Also, the FBO's are directed to obtain the Central License for their head office and if they operate in more than one state. The minimum tenure of this license is one year and the maximum is 5 years.

Documents Required for Basic FSSAI – Food License

Latest Passport sized photograph of the applicant, Identification proof (voter ID card or Aadhaar Card), PAN Card, Address Proof- Telephone or Mobile Bill/ Electricity or Gas Bill), Copy of Property papers (if owned), Copy of Rent Agreement and NOC letter from the Landlord (if rented)

Documents Required for FSSAI State License

Form-B duly completed and signed by the Proprietor, Blueprint or layout of the area location, Proof of possession of premises, Partnership affidavit of Proprietorship, Contact details of the Directors, List and details of the types of equipment and machinery, Proprietor's photo ID and Address Proof issued by the Government of India. In case of a company, the Directors photo ID and address proof, List of food category that has to be manufactured, Authority letter with name and address of the responsible person, NOC and Copy of License from the manufacturer and Food Safety Management System Plan or Certificate.

FSSAI License Registration Procedure

1. *Process for basic Registration*

- For licensing it is mandatory that the food business operator has an active email ID and a telephonic phone number.
- The business name of the respective food business operator must be spelled out correctly in the application form as the same name will appear on the license.
- After the successful submission of the application, a unique reference ID is issued that is to be used in all the further applications.
- The next and the last step is payment of the prescribed fees. This is to be done by taking out a print of the acknowledgment and the online application and make a demand draft of the specified amount.

2. *Procedure for State License Application*

- Before applying for a State FSSAI License it is important to check the eligibility criteria as mentioned by the Food Safety and Standards Authority of India.
- After successfully checking the eligibility criteria the next step is to furnish the application form for obtaining the State FSSAI License. Every data provided to the department must be correct and appropriate.
- Remember that the generic names must be mentioned for the products and not the brand names.

- All the products as mentioned in the license must be approved as per the FSS Act, 2006.
- At last, you will be asked for how many years you want to apply. You have to choose the appropriate period of time for which you wish to take the license.
- Post-filing the form, the prescribed fees have to pay either through an online portal or by submitting the demand draft at the FSSAI state office.

Suspension and Cancellation of Food License

Under the below-mentioned circumstances, the food license can either be canceled or suspended:

- Food poisoning outbreaks those are associated with the spread of the disease.
- Serious food complaints in cases where the consumer's safety is affected.
- Food Business Operator's non-compliant premises.
- Serious violations of the FSSAI rules and compliances.
- Non- compliance with an improvement or other legal notice without reasonable excuse.
- Interrupting an officer doing inspection.

Conclusion

Large number of innovative technologies and modern equipment's has been developed for preservation of fruits and vegetables. Linkages among farmers, entrepreneurs, investors and market are need of the hour for development of processing infrastructure in the country. Establishment of processing industry and value chain development by forward linkages with market and backward linkages with farmers will help in doubling farmer's income, employment generation and reduction in post-harvest losses.

Chapter 11

Importance and Scope of Processing of Horticultural Crops for Economic Upliftment of Farmers

Priya Awasthi, S.C. Singh and Er. H.M.Mandge

Banda University of Agriculture and Technology, Banda, Uttar Pradesh

The Indian economy is largely agrarian, with around 55 per cent of the population dependent for their livelihoods on agriculture and allied sectors that generate 15 per cent Gross Value Added (GVA) (GoI 2017a). Indian farmers are vulnerable to impacts of climate change, water scarcity and land degradation. In addition, increasing fragmentation of holdings, extreme weather events, rising input costs and post-harvest losses pose an enormous challenge to sustaining agricultural growth. Unfortunately about 25-30 per cent of horticulture produce, 10-25 per cent of vegetables and 30-40 per cent of flowers gets wasted due to lack of postharvest management which resulted in huge loss of crores of rupees.

The minimization of these postharvest losses may be reduced by extending the shelf life of fresh horticultural produces either through pre or post-harvest management practices or by processing it into different value added products. Several factors influence the post-harvest losses in fruits and vegetables that include losses due to physical, physiological, mechanical and unhygienic conditions, lack of proper storage conditions, refrigerated facilities and diseases and pests, *etc.* While harvesting to handling for storage till marketing several wound pathogens are known to infect the produced that destroy the keeping quality, quantity ultimately economic losses. Post-harvest decay of fruits and vegetables occur either between flowering and fruit maturity or during harvesting and subsequent handling and storage. There are many technologies already developed in the past which are available in the literature but are not practiced may be due to either materials are not available locally, not much effective or the technology is more costly. By adoption of simple post-harvest management practices, processing and value addition operation *viz.*, proper harvesting, sorting, grading, packaging, pulping, pickling, drying and dehydration at farmer's level during the peak season will help in minimization of post-harvest losses as well as doubling the farmer's income.

Fruit and vegetable processing industry has taken a new direction and is growing gradually with strong growth rate annually. Infrastructure development for processing may take this industry to the new heights in the years to come with the help of sufficient exports and investments. Fruit and vegetable processing is done widely in the food and beverage industry. Fruits and vegetables are processed into a variety of products such as juices and concentrates, pulp, canned and dehydrated products, jams and jellies, pickles and chutneys *etc.* The extent of processing of fruits and vegetables varies from one country to another. The fruit and vegetable processing industry in India is estimated to grow at a CAGR of 10.8 per cent between FY 2015 and FY 2020. The food processing sector is critical to India as it contributes to the economy, increases agricultural yields, creates employment and raises life-standards of people across the country, especially in rural areas.

The stage of crop production immediately following **harvest**, including cooling, cleaning, sorting and packing may be termed as Post Harvest Handling. Fresh Fruits and Vegetables including root crops are mostly perishable in nature. They begin to deteriorate immediately after its separation from the parent plant and suffer considerable losses during the process of Marketing. These losses may be of following nature:

1. Quality downgrading due to handling
2. Physiological spoilage (rooting and sprouting)
3. Pathological Spoilage from pest and diseases
4. Oversupply to markets

Important Sites of Post-harvest Losses

Important sites where post-harvest losses are noticed in India are:

- Farmer's field (15-20 per cent)
- Packaging (15-20 per cent)
- Transportation (30-40 per cent)
- Marketing (30-40 per cent)

On the Other hand, Value Addition in Horticulture is the process in which a high price is realized for the same volume of a primary product, by means of processing, packing, upgrading the quality or other such methods. For example: Making Jam of Pomegranate and Strawberry.

Need for Value Addition in Horticulture

- To improve the profitability of farmers.

- To empower the farmers and other weaker sections of society especially women through gainful employment opportunities and revitalize rural communities.
- To provide better quality, safe and branded foods to the consumers.
- To emphasize primary and secondary processing.
- To reduce post harvest losses.
- Reduction of import and meeting export demands.
- Way of increased foreign exchange.
- Encourage growth of subsidiary industries.
- Reduce the economic risk of marketing.
- Increase opportunities for smaller farms and companies through the development of markets.
- Diversify the economic base of rural communities.
- Overall, increase farmers' financial stability.

Importance of Value Addition in Horticulture

- Horticulture deals with variety of crops having great medicinal, nutritional, health promoting values.
- India as second largest producer of fruits and vegetables, only 10 per cent of that horticultural produce is processed, but other developed and developing countries where 40-80 per cent produce is value added.
- Horticultural crops provide varied type of components, which can be effectively and gainfully utilized for value addition like pigment, amino acids, oleoresins, antioxidants, flavors, aroma *etc.*
- Post harvest losses in horticultural produce are 5 to 30 per cent which amounts to more than 8000 crore rupees per annum. If we subject our produce to value addition the losses can be checked.
- Horticultural crops are right material for value addition because they are more profitable, has high degree of process ability and richness in health promoting compounds and higher potential for export.

Methods of Processing of Fruits into Products

- Preservation by heat treatment.
- Asceptic packaging.
- Preservation of by removal of heat.
- Quick freezing.
- Preservation by removal of moisture.
- Preservation by addition of chemicals.
- Minimal processing.

India is the major producer of dried and preserved vegetable like preserved onions, cucumber and gherkins, provisionally preserved mushrooms of the genus agaricus, other mushrooms and truffles, green pepper in brine, dehydrated garlic powder, dehydrated garlic flakes, Potatoes dried, grams dal, *etc.* Many non-traditional vegetables mainly processed cucumber and gherkins and other vegetables produced like asparagus, celery, bell pepper, sweet corn, green and lime beans and organically grown vegetables are also being increasingly exported. It is envisaged that India should be the food factory of the world so as to generate income, employment and foreign exchange in a big way, through significant increase in level of processing and achieve higher growth through value addition, increasing India's share in global and achieve quality image for Indian products in domestic as well as overseas market.

Significant developments in technology include better understanding of the process of ripening of fruits, optimum harvesting time, pre-cooling of freshly harvested produce, cold storing of the raw fruits and vegetables, sorting, cleaning, waxing, packaging technology for fruits. At CFTRI, DFRL, IIHR, Bangalore; IARI, New Delhi; GBPUA&T, Pantnagar; IIVR, Varanasi and HPKV, Palampur; a number of technologies have been developed. Most significant work has been recorded in the technology for ripening of the fruits under controlled conditions. Production of juices and value-added products including jams, jellies, pickles, canned products *etc.* has become a commercial success.

The industry using indigenous technology includes units engaged in juice extraction, concentration of juices, canning and production of several of the products like jams, jellies, canned fruits, dried vegetables *etc.* Technology is still being imported for establishment of large scale exported oriented units for production of items like banana paste, concentrates of various fruit juices, sorting, cleaning, washing, waxing and packaging of raw fruits and vegetables.The production of fruit and vegetable products in India are canned, bottled fruits and vegetables, jams, jellies, marmalades, fruit juices, fruit pulps, squashes, crashes, cordials, fruit syrups, fruit nectars, RTS fruit beverages, fruit juice concentrates, chutneys, pickles, mango slices in brine preserves, candied and crystallized fruits and peels, dehydrated fruits and vegetables, frozen fruits and vegetables, tomato products, sauces, soups *etc.*

References

A.K. Thompson 2010. Controlled Atmosphere Storage of Fruits and Vegetables, CABI Publishing; 2nd revised edition.

Alzamora, S.E., Tapia, M.S., and López-Malo, A. 2000. Minimally Processed Fruits and Vegetables: Fundamental Aspect and Applications. Aspen Pub. Co., Inc., Maryland, US, 277-286.

Alzamora, S.M., Cerruti, P., Guerrero, S., and López-Malo, A., Díaz, R.V. 1995. Minimally processed fruits by combined methods. In: Fundamentals and Applications of Food Preservation by Moisture Control. ISOPOW Practicum II, Barbosa-Cánovas, G. and Welti, J., Eds., Lancaster, PA, Technomic Publishing, 565-602.

Arthey, D. and Ashurst, P.R. 1996. Fruit Processing. Blackie Academic and Professional, London, 142-125.

Barbosa-Cánovas, G.V. and Vega-Mercado, H. 1996. Dehydration of Foods. Chapman and Hall, New York, 53-59.

Bhattachajee SK and Dee LC. 2005.Post harvest technology of flowers and ornamental plants. Pointer Publishers, Jaipur.

Cantwell, M. 2001. Post-harvest Handling Systems: Minimally Processed Fruit and Vegetables. Vegetable Information. University of California, Vegetable Research and Information Center.

Government of India. 2017A. Agricultural statistics at a glance 2016. Government of India, Ministry of Agriculture, Department of Agriculture and Cooperation, Directorate of Economics and Statistics.

Gupta, Ajay.2016.List of Profitable Value Added Products from Fruits and Vegetables. Value Addition and Processing of Agri-Products. www.entrepreneurindia.co, www.niir.org

L.R. Verma, V.K. Joshi. 2000.Post Harvest Technology of Fruits and Vegetables. Vol. I and II. Vegetables. Indus Publishing Co. New Delhi.

Stanley J. Kays. 1998. Post Harvest Physiology of Perishable Plant Products.CBS New Delhi.

Wills, McGlasson, Graham Joyce.2007. Post Harvest- An Introduction to the Physiology and Handling of Fruits, Vegetables and ornamentals. Cab International.

Sanal, B. and Krishna Kumar. S. 2017.An analysis of value addition in agro products and its impact on the export potentials of India. *International Journal of Management*, Volume 8, Issue 4, July–August 2017, pp.23–30.

Chapter 12

Nanotechnology for Augmenting Farmers' Income

Rajeew Kumar[1]*, Jagannath Pathak*[2]*,*
Priyanka Pandey[3] *and Nikhil P Singh*[4]

[1]*Senior Research Officer–Agronomy,*
[3]*Assistant Professor –MBGE,*
GB Pant University of Agriculture and Technology, Pantnagar
[2]*Associate Professor –Soil Science and Agril. Chem.,*
Banda University of Agriculture and Technology, Banda, Uttar Pradesh
[4]*Ph.D Scholar, Venketashwra University, Gajraula*

Food security is one of the great global challenges of the 21st century that will require solutions on the time scale of only a few decades. The global population grew to seven billion in 2011, and by 2050 it is expected to reach over nine billion. However, the resources to feed this growing population are limited. In 1960, one hectare of land produced enough food to feed two people; by 2050, the same amount of land will need to feed more than six people (FAO). Soils, and their continuing ability to support the sustainable intensification of agriculture, will play a central and critical role in delivering food security. The pressure on soils is urgent and growing, but the complexity of soils creates significant challenges to establishing the robust science base needed to support key decisions on their future management. Since soils and food supplies are global issues, gaining and maintaining international consensus on soil management poses further scientific and diplomatic challenges. The pressure on soils is significant and growing. A concerted response is required from researchers, politicians and land users across a broad range of research disciplines, jurisdictions and commercial markets, to ensure that soils are protected and enhanced. So, to feed a growing population under increasing natural resource constraints, organizations are promoting "sustainable intensification" as the future of agricultural production (FAO, 2009). The application of nano-technology techniques to agricultural crop inputs is one of the proposed tools for "sustainable intensification."2 These appli-cations include reducing the volume of pesticide use through adding nano-silver particles to pesticides to make them more effective in targeting pests with a smaller pesticide volume; adding nano–metal oxides to target soil pathogens, *e.g.*, those resulting from fertilizing with non-composted manure; adding nano-silicon to increase water uptake efficiency in plants; developing a DNA-based nanobio-sensor in a polymer to coat fertilizers, which would release only as much fertilizer as "demanded" by plant root ionic signals. Rhizosphere (root interaction area) is the main attraction within the soil as it is directly responsible for crop production. Newman (1985) examined a variety of plant species and estimated that roots can release anywhere from 10 to 250 mg C/g root produced or about 10-40 per cent of their total photosynthetically fixed carbon. The C released is in both organic (*e.g.*, low molecular weight organic acids) and inorganic (*e.g.*, HCO_3) forms, however, the organic forms are the most varied and can have the most influence on the chemical, physical and biological processes in the rhizosphere (Jones, *et al.*, 2009). The composition and amount of the released compounds is influenced by many factors including plant type, climactic conditions, insect herbivory, nutrient deficiency or toxicity, and the chemical, physical and biological properties of the surrounding soil. Roots exert a tremendous amount of pressure (>7kg/cm^2 or ~100 psi) at the growing root tips in order to push their way through the soil. Helping to lubricate and protect the root during growth, root cap and epidermal cells secrete mucilage, a viscous, high molecular weight, insoluble, polysaccharide-rich material. Beyond lubrication, the mucilage also provides protection from desiccation, assists in nutrient acquisition, and most notably binds soil particles together forming aggregates which improve soil quality by increasing water infiltration and aeration. Also serving to protect the root tip, cells lining and capping the root meristem are programmed to release (slough-off) which helps to reduce frictional forces that would otherwise damage the root (Bengough and McKenzie, 1997).

Nanotechnology

Importance of agriculture to all human societies is characterized more than ever with increasing world population. The first and most important need of every human is needs to food, and food supply for humans associated with agriculture directly and indirectly. Growth of the agricultural sector as a context for development objectives is seen as essential in developing countries. Now, after years of green revolution and decline in the agricultural products ratio to world population growth, it is obvious the necessity of employing new technologies in the agriculture industry more than ever. Modern technologies such as bio and nanotechnologies can play an important role in increasing production and improving the quality of food produced by farmers. Many believe that modern technologies will secure growing world food needs as well as deliver a huge range of environmental, health and economic advantages. Food security has always been the biggest concern of the mankind. Nations, communities and governments have been struggling with the issue since long. Recent decades have seen even bigger challenges on this front. The future looks even bleaker with food shortage issue looming large. The challenge is how to feed the growing population by producing more on a stagnant or shrinking landscape; with lesser input costs and with lesser hazards to the eco-system.

Nanotechnology is an interdisciplinary field that has been entered in different range of applied sciences such as chemists, physicists, biologists, medical doctors and engineers. Targeted research and development, for understand, manipulate and measure at the materials with atomic, molecular and super molecules dimensions is called nanotechnology. Nanotechnology has been provisionally defined as relating to materials, systems and processes which operate at a scale of 100 nanometers (nm) or less. A nanometer is one billionth of a meter. Overall nano refers to a size scale between 1 nanometer (nm) and 100 nm. For comparison, the wavelength of visible light is between 400 nm and 700 nm. A leukocyte has the size of 10000 nm, a bacteria 1000-10000 nm, virus 75-100 nm, protein 5-50 nm, deoxyribonucleic acid (DNA) ~2 nm (width), and an atom ~0.1 nm In this scale, physical, biological and chemical characteristics of materials have fundamentally different from each other and often unexpected actions are seen from them. Nanotechnology considers the topics with viruses and other pathogens scale. So has high potential for identify and eliminate pathogens (Rohella 2011).

Historically the exact time period from which the application of nano-particles started is not known but available evidences indicate that gold nano-particles were used in glazes in ancient pottery to produce different colored glazes. During the medieval period, extensive use of different gold nano-particles were done and were incorporated into the glass to impact different colors. Such colored glasses were used in medieval and Victorial churches. However, those glassmaker or the glaziers did not know how and why different sized gold particles exhibited different colors. This vision was reflected in his lecture entitled "There's plenty of Room at the Bottom", delivered at the annual meeting of the American Physical Society at Callech in 1959. A Japanese Scientist Norio Taniguchi at the University of Tokyo first used the term "Nano-technology" in 1974 to refer to the ability to engineer materials precisely at the nano-meter level. The dreams of Richard Feynman came to reality with the great research worker of Eric Drexler during Late 1970s and he started publishing scientific papers from 1981. In 1986, he introduced the term "nano-technology" in his book "Engineer of Creation" to describe his approach of molecular manufacturing and some of its consequences. The nano-science research took a new momentum with the invention of Scanning Tonnelling Microscope (STM) in 1981-82 by Greg Karl Binning and Heinrich Rohren and Atomic Force Microscope (AFM) in 1986. A direct stimulus to the systemic study of nano-science came from the discovery of fullerenes in 1985 Harold Kroto, Robert F. Curl and Richard Smalley for which they were jointly awarded Nobel Prize in chemistry in 1996. Earlier we used to know three forums of Carbon, namely, diamond, graphite and amorphous carbon. Today, there is a whole family of other forum of carbon comprised of hollow cage like molecular, popularly known as Fullerenes, comprised of 60 to 120 atom of carbon. Popularly, C60 molecular of carbon is referred to as buckyball, in honour of Buckminniter Fuller.

Nanotechnology development in India is at a nascent stage with policy initiatives directed towards promoting research and development. It is largely a government lead initiative and industry participation is still emerging (Annonymus 2009). A more inclusive science and technology (S&T) policymaking process and integrated framework is needed to ensure that this technology is diffused and used for socially valued goods. At present there are more than 30 institutions in the country that are involved in Research and Development and training programmes in this area. The Indian government has spent Rs. 1000 million for nanotechnology under its 10 century architect renowned for his use of geodesic domes. This led in turn, to the discovery of "Carbon nano-tubes" (CNTs) in 1991 by the Japanese electron microscopist Sumio Lijima. These nano-tubes are about 100 times stronger than steel but just a sixth of the weight and they have unusual heat and conductivity characteristics. In general, precision agriculture is a new attitude in farm management. With use of nano sensors will be determined every small part of farm how much needs to fertilizer and chemical pesticides. Therefore, use of inputs will be optimal and safe products and economic efficiency is increased. Nano-sensors help to farmers in maintaining farm with precise control and report timely needs of plants. Nano sensors and nano-based smart delivery systems could help in the efficient use of agricultural natural resources like water, nutrients and chemicals through precision farming. Through the use of nano materials and global positioning systems with satellite imaging of fields, farm managers could remotely detect crop pests or evidence of stress such as drought. Once pest or drought is detected, there would be automatic adjustment of pesticide applications or irrigation levels. Nano sensors dispersed in the field can also detect the presence of plant viruses and the level of soil nutrients. Nano fertilizers will be absorbed by plants rapidly and completely. Nano encapsulated slow release fertilizers have also become a trend to save fertilizer consumption and to minimize environmental pollution. Super water adsorbents made by nanotechnology, has an important role in storage and protecting water in arid and semiarid regions. Nanotechnology has many applications in the field of agricultural machinery such as: application in machines structure and agriculture tools to increase their resistance against wear and corrosion and ultraviolet rays; producing strong mechanical components with use of nano-coating and use of bio-sensors in smart machines for mechanical-chemical weed control.

References

Anonymous, 2009. Nanotechnology and nano science applications: revolution in India and beyond. Strategic

Bengough, A.G. and McKenzie, B.M. Sloughing of root cap cells decreases the frictional resistance to maize (*Zea mays* L.) root growth. *J. Exper. Bot*, Vol. **48**(309):885-893 (1997).

Food and Agriculture Organisation of the United Nations, Farming Must Change to Feed the World, http://www.fao.org/news/story/en/item/9962/icode/

Food and Agriculture Organization, 2009. Available at http://www.fao.org/agriculture/crops/core-themes/theme/spi/scpi-home/framework/sustainable-intensification-in-fao/en/.

Sayed Roholla Mousavi; Maryam Rezaei. 2011. Nanotechnology in Agriculture and Food Production. *J. Appl. Environ. Biol. Sci.*, 1(10)414-419, 2011

Chapter 13

Precision Agriculture: A Way to Enhance Farmers Income

Rajeew Kumar[1], *Jaganath Pathak*[2], *Priyanka Pandey*[3] *and Nikhil P. Singh*[4]

[1]*Senior Research Officer–Agronomy,*
[3]*Assistant Professor–MBGE,*
GB Pant University of Agriculture and Technology, Pantnagar
[2]*Associate Professor–Soil Science,*
Banda University of Agriculture and Technology, Banda
[4]*Ph.D Scholar, Venketashwra University, Gajraula*

Agricultural sustainability in India can be achieved only if the natural resource base upon which it relies is well managed. However, in attempt to achieve high productivity over past 3-4 decades, a critical linkage between agriculture and the environment fragility has been ignored. Even intensively managed cropping systems in India seem to be unsustainable as yield levels have become stagnant or declined with a decreasing response to farm inputs and a widening gap between the potential and realized yields. Rapid increase in population pressure, urbanization and income growth are now exerting extra pressure to further increase production. Hence, for developing country like India, adoption of environment-friendly technologies and holistic approach of farm management is, therefore, essential for agricultural decision making. Modern frontier technologies involving system approach towards efficient crop and input management, and scientific land and water use planning, is thereby the need of this century sustainable agricultural management.

Precision Agriculture

Precision agriculture (PA) – also known as precision farming, prescription farming, variable rate technology (VRT) and site specific agriculture – is a current buzz word among the agricultural circles and considered as the agricultural system of the 21st century, as it symbolizes a better balance between reliance on traditional knowledge and information – and management – intensive technologies. It is an integrated agricultural management strategy where farmers can adjust input use and cultivation methods – including seed, fertilizer, pesticide, and water application, varietal selection, planting, tillage, harvesting – according to varying soil, crop and other field conditions. In brief, precision agriculture refers to tailoring crop and soil management practices according to variation in crop and soil conditions within each field. Precision agriculture technologies provide three basic requirements for precise and sustainable agricultural management. These are: Ability to identify precise location of field, 2. Ability to gather and analysis information on spatio – temporal variability of soil and crop conditions at field level, and 3. Ability to adjust input use and farming practices to maximize benefits from each field location.

Role of Precisiton Agriculture

The real value of precision agriculture for the farmer is that he can adjust seeding rates, plan more accurate crop protection programs, perform more timely tillage and know the yield variation within a field. These benefits will enhance the overall cost effectiveness of his crop production.

Seeding

Hybrid seeds perform best when placed at spacing that allow the plants to obtain such benefits as maximum sunlight and moisture. This is best accomplished by varying the seeding rate according to the soil conditions such as texture, organic matter and available soil moisture. One would plnat fewer seeds in sandy soil as compared to silt loam soils because of less available moisture. The lower seed population usually has larger heads (ears) of harvested seeds providing for a maximum yield. Since soils vary even across an individual farm field, the ability to change seeding rates as one goes across the field allows the farmer to maximize this seeding rate according to the soil conditions. A computerized soil map of a field on a computer fitted on the tractor along with a GPS can tell farmers where they are in the field allowing the opportunity to adjust this seeding rate as they go across their fields.

Crop Protection

The application of chemicals and fertilizers in proper proportions are of environmental and economic concern to the farmers. Environmental regulations are calling for the discontinuance of certain pesticide applications within 100 feet of a stream or water body or well or within 60 feet of an intermittent stream. Using a GPS along with a digital drainage map, the farmer is able to apply these pesticides in a safer manner. In fact, the spraying equipment can be preprogrammed to automatically turn off when it reaches the distance limitation or zone of the drainage feature. Additionally, farmers can preprogram the rate of pesticide of fertilize to be applied so that only the amount needed determined by the soil condition is applied varying this rate from one area of th field to another. This saves money and allows for safer use of these materials.

Tillage

The ability to vary the depth of tillage along with soil conditions is very important to proper seedbed preparation, control of weeds and fuel consumption and therefore cost to the farmer. Most farmers are using conservation tillage which means leaving residues on the soil surface for erosion control. The use of GPS in making equipment adjustments as one goes across the different soil types would mean higher yields and safer production at lower costs. This part of precision farming is in its infancy. The equipment companies will be announcing tillage equipment with GPS and selected controls tailored to precision farming in the near future.

Harvesting

The proof in the use of variable rate technology (adjusting seed, pesticide, fertilizer and tillage) as one goes across the field is in knowing the precise yields. Combines and other harvesting equipment can be equipped with weighing devices that are coupled to a GPS. One literally measures yields on the go. With appropriate software, a yield map is produced showing the yield variation throughout the field. This allows farmers to inspect the precise location of the highest and the lowest yielding areas of the field and determine what caused the yield difference. It allows one to program cost and yield to determine the most profitable practices and rates that apply to each field location. In my opinion, the use of yield monitors is a good place to start if one wants to get started in precision farming. Yield data from the same field over 3 + years would define the weak spots in the field and narrow down the probability of what is causing a log yield.

Scope of Precision Agriculture in India

Precision farming technologies is a suite of many high-tech tools but there is no need to adopt all PF technologies at once to start benefiting from them. Many farmers can begin by using only a part of the technology, as even partial use can bring many benefits. In fact, applying the entire range of technologies is not profitable in several cases, particularly for technologies that are not scale-neutral. For example, small farmers in India cannot afford on their own, but some private sector support is needed for the advancement of data acquisition and analysis methods, including sensing technologies, sampling methods, data base systems, and geospatial methods. Some of the agribusiness companies like Nagarjun Fertilizers Company Limited, BAYER India Ltd. and Mahyco Seeds Pvt. Limited should come forward and get actively involved in extending the services on precision farming technologies to the farmers someway. There are many companies have involved in these extension activities and helping to the farmers.

Precision farming technologies is likely to provide a greater profitability advantage for (a) high-value crops, (b) areas where input costs are high, and (c) areas where production conditions are very heterogeneous. The implementation of precision agriculture technologies in India should have two different strategies – one for the low input subsistence agriculture and the other for input-intensive profit-making agriculture. In case of former, the increase in productivity is the prime concern. Here, the system has to be converted to information-based agriculture, where farmer has spatial information about the soil and crop. This information can be used for efficient input application. Since the field size are small in this situation individually bunded field or a group of field can be considered as a unit for variable rate application. However, for the latter case, such as rice and wheat of Indo-Gangetic belt and the horticultural crops like grape (Maharashtra), potato (Punjab), tea (Assam) where the field size arer large and farmers are rich, input use efficiency is prime concern. Her, remote sensing data can be used to identify the spatial and temporal variability and necessary actions can be adopted using sophisticated equipments like varable rate technology. Adoption of PF techniques aimed at irrigation management, nutrient management and integrated pest management will obviously be a priority for such crops.

Limitaions for Adoption of Precision Agricultute in India

There are many limitations in adopting this high-tech precision farming technology in India. Some of them include,

- High cost of obtaining site-specific satellite data
- Lack of willingness to share spatial data among various organizations
- Complexity of tools and techniques requiring new skills
- Culture, attitude and perceptions of farmers including resistance to adoption of new techniques and lack of awareness of agro-environmental problems
- Farmers inability to afford High-tech farm equipments
- Small farms, heterogeneity of cropping systems, and land tenure/ ownership restrictions
- Infrastructure and institutional constraints including market imperfections
- Lack of success stories of PE adoption and lack of demonstrated impacts on yields
- Lack of local technical expertise
- Uncertainty on returns from investments to be made on new equipment and information management system
- Lack of transformation of technical know –how to farmers in local language, and
- Knowledge and technological gaps including
- Inadequate understanding of agronomic factors and their interaction.
- Lack of understanding of the geostatistics necessary for displaying spatial variability of crops and soils using current mapping software, and
- Limited ability to integrate information from diverse sources with varying resolutions and intensities.

References

Ravi, N. and Jagadeesha, C.J. 2002. Precision Agriculture, Training course on Remote Sensing and GIS Applications in Agriculture, May 27th –7th June, 2002, RRSSC-Bangalore, pp: 225-228.

U.K. Shanwad, V.C. Patil, G. S. Dasog, C.P. Mansur and K. C. Shashidhar. 2002. Global Positioning System (GPS) in Precision Agriculture the Asian GPS Conference 2002, (October 24 – 25, 2002, New Delhi, India).

Chapter 14

Onion and Garlic: Excellent Potential for Enhancing Profitability of Farmer's in Bundelkhand Region of UP

R.K. Singh, A.C. Mishra, S.V. Diwedi, S. Kumar, Neetu and Sweta Soni

Department of Vegetable Science, College of Horticulture Banda University of Agriculture and Technology, Banda – 210 001, U.P.

Vegetable crops play a pivotal role in food and nutritional security of the growing population and these occupy an important place in crop diversification leading to higher income by combating ill-effects of climate change in agricultural production system in India. India is the second largest producer of fruits and vegetables in the world. Onion is known to the mankind since prehistoric times. Its reference is mentioned in the religious books like Bible, Koran and inscriptions of ancient civilizations of Egypt, Greece and China. The Greeks and Romans were using onion since 400-300 BC. In India, it is known to have been cultivated since 600 BC. Total area under horticultural crops is 21.83 million hectares and production are 240.53 million tones. Fruits and vegetables together contribute about 92 per cent to the total horticultural production in the country. Productivity of vegetables in India (17.3t/ha) is less than the world average productivity (18.8t/ha). Onion and garlic are widely cultivated for internal consumption as well as for the export.

The annual production of onion is **21.83 million** tons from **1.31 million ha** area with a productivity of **16.60 tones/ha**. However, the annual production of garlic are 1.71 million tons from 0.33 million tones hectare with a productivity of 5.16 tones/

ha. Both are used as spices, condiments and vegetables almost daily in every kitchen as a seasoning for wide varieties of dishes. The green leaves, immature and mature bulbs are eaten raw as well as cooked vegetables. Because of its special characteristics of pungency, it has more value than other vegetables. By the export of onion and garlic, country earns a significant amount of foreign exchange. Day by day export demand for these crops is increasing. In north Indian conditions, onion and garlic is generally grown during *Rabi* season. However, in South India, three crops *viz.*, *kharif*, late *kharif* and *Rabi* crops are taken.

Onions are used as spices, condiments and vegetables almost daily in every kitchen as a seasoning for wide varieties of dishes. Because of its special characteristics of pungency, it has more value than other vegetables. Onion is used as salad or cooked in various ways in all curries, fried or baked, boiled, soup-making and processed forms, *e.g.* flakes, powder paste, crushed onion and pickles. The dehydrated flakes or powder, a part of fresh onion, is in great demand in domestic and export markets. The pungency of onion is due to the presence of Sulphur compound. The main compound is allyl-propyl disulphide.

The Bundelkhand region of Uttar Pradesh comprising Chitrakoot, Banda, Hamirpur, Mahoba, Lalitpur, Jalaun and Jhansi districts having 3.2 crores population is climatically most challenged and drought prone area. Due to declining and erratic rainfall, Bundelkhand region frequently suffers drought. This region has very good potential for *Rabi* and *Kharif* onions. The farmers of this area generally grow local varieties of onion and garlic in *Rabi* season with reasonably very low yield. In many parts of the country, particularly, in Deccan plateaus *kharif* onion is grown with quite higher productivity. As the Bundelkhand region receives comparatively low rainfall during *Monsoon*, possibility of *kharif* onion in this region is equally high. It is a very good advantage of this region is the soil of this region is rich in Sulphur content, and onion and garlic are loving plant of sulphur content, therefore, it is much possibility to cultivate the onion and garlic to enhance the production as well as productivity of both the crops in this region. However, intensive research work is required in relation to evaluation of high yielding genotypes/varieties and hybrids, soil amendments, standardization of nutrient doses and other cultivation practices.

A number of varieties of onions and garlic have been developed in India by different SAUs and ICAR Institutes such as DOGR, Rajgurunagar, IIHR, Bengaluru, IARI, New Delhi, MPKV, Rahuri, PAU, Ludhiana, TNAU, Coimbatore and NHRDF, Nasik for cultivation in different regions in the country. Some of the important cultivars under commercial cultivation in different parts of the country.

Therefor it is a need to develop varieties suitable for different seasons for Bundelkhand region for more adoptability, having resistance to biotic and abiotic stresses, suitable for processing industry, green foliage, export quality, mechanized farming for large as well as small farmers, having better keeping quality for enhancing the farmers income of Bundelkhand region of Uttar Pradesh.

Global Perspective

Onion is widely grown in the world covering an area of 4.44 million hectares with a total production of 85.80 million tones and productivity of 19.31 tones/ha.

Garlic area in the world is 1.44 million hectares with a total production of 24.32 million tones and productivity is 16.85 tones/ha. China and India are the leading onion and garlic growing countries with a total production of 22.34 and 19.30 million tones in onion and 19.23 and 1.26 million tons of garlic respectively. Other major onion and garlic producing countries are USA, Egypt, Iran, Turkey, Brazil, Pakistan, Russian Federation, Republic of Korea and Netherlands. In the major onion producing countries, productivity is highest (64.58 tons/ha) in Republic of Korea, followed by USA (54.47 tons/ha), Spain (53.69), Netherlands (45.80 tons/ha) and Japan (42.46 tons/ha). The productivity of garlic amongst the major producing countries is highest (39.49 t/ha) in Uzbekistan followed by Egypt (25.17 t/ha), China (24.56 t/ha) and USA (18.14 t/ha). India even being a major onion and garlic producing country has quite low productivity of 15.86 and 5.08 tons/ha respectively.

History of Research and Development of Onion and Garlic

The Indian Agricultural Research Institute (IARI), New Delhi, and Indian Institute of Horticultural Research (IIHR), Bengaluru and SAUs initiated onion breeding programmers. The NHRDF, Nashik developed several onion varieties and technologies for onion production. Development of multiplier onion varieties was done by Tamil Nadu Agricultural University (TNAU), Coimbatore. Prior to this, research on collection and maintenance of land races and standardization of agro-techniques was attempted by various State Agricultural Departments. With the concept of coordinated projects and initiatives of SAUs onion research was strengthened, in terms of varietal development for different seasons and standardization of production techniques in early nineties. The systematic R&D activities on garlic crop started after establishment of National Horticultural Research and Development Foundation during 1977. Later on work was also initiated by different State Agricultural Universities and ICAR Institutes. Major work on development of varieties and other aspect are being taken up at NHRDF, DOGR Rajgurunagar, Central Institute of Tropical Horticulture Srinagar, Jammu and Kashmir, HAU Hissar, MPKV Rahuri, PAU Ludhiana, VPKAS Almora, ARU Almora, IARI, New Delhi, GAU Junagarh in the country. The work in the garlic research was strengthened in all aspect especially on varietal development, production and post-harvest technology after starting the All India Coordinated Research Project on vegetable crops in early nineties. The R&D in onion and garlic got impetus with the establishment of National Research Center on Onion and Garlic at Nashik in 1994.

A large number of onion and garlic varieties have been developed by different Universities and Institutes in India for cultivation in different regions and also in different seasons.

Common Big Onion

Red Coloured Bulbs

Punjab Selection, Pusa Ratnar, Pusa Red, N-2-4-1, Pusa Madhavi, Arka Niketan, Arka Kalyan, Agrifound Dark Red, Kalyanpur Red Round, Agrifound Light Red, Hissar-2, Punjab Red Round, Arka Pragati, N-53, Baswant -780, Udaipur-101,

Udaipur -103, Phule Samart,h NHRDF Red (L-28), NHRDF Red-2 (L-355), NHRDF Red-3 (L-652,) NHRDF Red-4 (L-744), Bheema Super, Bheema Raj, Bheema Red, Bheema Shakti, Bheema Kiran and VL–Piaz-3

Yellow Coloured Bulbs

Early Grano, Spanish Brown, Phule Suvarna and Arka Pitamber

White Coloured Bulbs

Pusa White Round, Pusa White Flat, Punjab-48, Udaipur-102, Phule Safed, N-257-9-1, Agrifound White, Akola Safed, JNDWO, Bheema Subra

Small Common Onion

Agrifound Rose, Arka Bindu, Arka Swatista

Multiplier onion

Co-1, Co-2, Co-3, Co-4, MDU-1, Agrifound Red, Co-On-5.

Arka Ujjwal

Onion Hybrids

Arka Kirtiman, Arka Lalime, Hybrid – 63, Hybrid - 35

Table 14.1: Garlic Varieties Developed by different Organization

Organization	*Variety*	*Colour*	*Zones*
National Horticultural Research and Developed Foundation, Nashik	Agrifound White	White	IV, VII
	Yamuna Safed	White	IV, VI, VII
	Yamuna Safed-2	White	IV
	Yamuna Safed-3	Creamy white	IV,VI,VII
	Yamuna Safed-4	White	IV, VI,VII
	Yamuna Safed-5	White	III, IV,VI
	Yamuna Safed-8	White	II
	Yamuna Safed-9	Purple	II
	Agrifound Parvati	Purple White	I
	Agrifound Parvati-2	White	I
Mahatma Phule Krishi Vidyapeeth, Rahuri, Maharashtra	Godavari	Purple	VII
	Sweta	White	VII
	GG 4	White	VI
	Phule Baswant	White	VII
Chaudhary Charan Singh Haryana Agricultural University, Hissar,	HG-1	White	VI
	HG-2	White	VI
Punjab Agricultural University Ludhiana Punjab	LCG-1	White	IV
Vivekanand Parvatiya Krishi Anusandhan Shala, Almora, Uttarakhand VPKS, Almora VPKS, Almora	VL Garlic 1	White	I
	VL Garlic 2	White	I
	VL-Garlic 7	Purple white	I

Organization	Variety	Colour	Zones
Tamil Nadu Agricultural University, Coimbatore, Tamil Nadu	Ooty 1	Dull white	VIII
Directorate of Onion and Garlic Research Rajgurunagar, Pune	Bhima Omkar	White	V
	Bhima Purple	Purple	V
Junagarh Agricultural University, Junagarh	GAUG-1	White	IV
	GAUG-10	White	IV
	GG-2	White	IV
Govind Balabh Pant University of Agriculture and Technology, Pantnagar	Pant Lohit	White	IV
Central Institute of Temperate Horticulture, Srinagar, Jammu and Kashmir	CITH-G-1	White Purple	I
	CITH-G-2	Purple	I

Important Onion and Garlic Varieties and their Description

Agrifound Dark Red

Selection made by National Horticultural Research and Development Foundation, Nashik. Variety was notified by Government of India in 1988 vide notification number 1135 (E) dated 01/12/1988 for the plains throughout the country, where *Kharif* onion is grown. Bulbs are dark red in colour, globular in shape with tight skin with moderately pungent. Bulb diameter is 4-6 cm. Total soluble solid 12-13 per cent and Pyruvic acid 10.07 micro mol/g. It is also suitable for bulblets production for early *Kharif* crop. Mature in 90-95 days after transplanting. Average yield 300 q/ha, keeping quality average and recommended for *Kharif* season all over the country.

Line-883

The plant is straight with erect green leaves and number of leaves per plant is about 6-9. Dark Red bulb colour, globular round shape, thin neck and 5.0-5.75 cm, bulb diameter. The average weight of 20 bulbs is about 1.00-1.35 kg. Total soluble solid and dry matter content is about 12.5-13 per cent and 13-14 per cent, respectively. The variety is accepted by farmers because of its higher yield and better adaptability. Mature in 85-90 after transplanting. In case of broadcasting of direct seeds in fields in Gujarat, Karnataka and Madhya Pradesh areas it matured in only 83 days with good and compact bulbs. Tolerant to foliar diseases likes purple blotch and stemphylium blight under field condition. Average yield is about 25-32.5 t/ha.

Agrifound Light Red

Selection made by National Horticultural Research and Development Foundation, Nashik through mass selection. Variety was notified by Government of India in 1996 vide notification number 115 (E) dated 10/02/1996 for Western, Northern Central and Southern India. Bulbs are globular in shape with tight skin, light red colour and 4-6 cm in diameter. Total soluble solid 13-14 per cent and Pyruvic acid 12.20 micro mole/g. Mature in 100-110 days after transplanting and keeping quality good. Average yield 300-325 q/ha and recommended for *Rabi* all over the country. It can be grown in late Kharif season also in Nashik area of Maharashtra.

NHRDF-Red

Selection made by National Horticultural Research and Development Foundation. Variety was notified by Government of India in 2006 vide notification number SO-2035 dated 28/10/2006 for Northern, Central and Western India. Bulbs are attractive very dark red in colour and 4.5-6 cm diameter. Total soluble solid is 13-14 per cent. Variety is very popular among farmers because of its higher yield, attractive dark red colour of bulbs and better storage performance. Mature in 110-120 days after transplanting. Average yield 250-300 q/ha. Recommend for cultivation in Northern, Central and Western India in *Rabi* season.

NHRDF Red-2

Selection made by National Horticultural Research and Development Foundation, Nashik through mass selection. Variety was notified by Government of India in 2012 vide notification number SO-2363 dated 04/10/2012. The plant straight, erect leaves, light red bulbs colour, globular round shape, thin neck and 4.0-6.0 cm in diameter. Total soluble solid 13-14 per cent and Pyruvic acid 12.0 micro mol/g. Mature in 100-120 days after transplanting and keeping quality good. Moderately resistant to foliar diseases like stemphylium blight. Average yield 350-400 q/ha and recommended for *Rabi* for III, IV and VI zones.

NHRDF Red-3

Selection made by National Horticultural Research and Development Foundation, Nashik through mass selection. The plant straight, erect leaves, bulb diameter-5.50-6.50 cm, colour of bulb-bronze red, shape of bulb globular round and tight skin. The total soluble solid ranges between 12.0-14.0 per cent and Pyruvic acid 12-13 micro mol/g. moderately resistant to foliar diseases like stemphylium blight. Average yield 350-450 q/ha with good keeping quality.

NHRDF Red-4

Onion variety, NHRDF Red-4 (L-744), developed by the NHRDF, Nashik, has been notified through Crop Standard Notification and Release of Varieties of Horticulture Crops, Ministry of Agriculture Cooperation and Farmer welfare, Government of India vide notification number SO-3666 (E) dated 06/12/2016 and released for cultivation in *rabi* season in zones III (Uttar Pradesh, Bihar, West Bengal and Imphal) the country **(2016)**. The plant straight, erect leaves, bulb diameter-5.50-6.25 cm, colour of bulb is dark red, shape of bulb is globular round with tight skin. The total soluble solid ranges between 12-14 per cent and Pyruvic acid 12-13 micro mol/g. Average yield 350-400 q/ha with good keeping quality.

Bhima Dark Red

Developed by DOGR, Rajgurunagar, Pune in 2012. Bulbs are attractive dark red globular bulb; bulbs mature 95-100 DAT. Total soluble solid (11-12 per cent). Average yield (200-220 q/ha) and recommended for *Kharif* season **in** zones II, IV, V and VI

Bhima Kiran

Developed by DOGR, Rajgurunagar, Pune in 2010. Bulbs are attractive red with round shape; bulbs mature 125-130 DAT. Total soluble solid (11-12 per cent). Average yield (280-320 q/ha). Recommended for *rabi* season **in** zones III and VI

Bhima Shakti

Developed by DOGR, Rajgurunagar, Pune in 2011, Bulbs are attractive red with round shape, bulbs mature 125-135 DAT, Total soluble solid (12-13 per cent). Average yield (280-300 q/ha) and recommended for late-*Kharif* and *rabi* season **in** zones III, IV, V and VI

Bhima Raj

Developed by DOGR, Rajgurunagar, Pune in 2007. Bulbs dark red, oval in shape and thin neck, bulbs mature 120-125 DAT. Keeping quality is average and total soluble solid (10.0-11.0 per cent). Average yield (250-300 q/ha) and recommended for *Kharif* season in zones VI, VII and VIII

Bhima Shweta

This white onion variety is already recommended for *Rabi* in Andhra Pradesh, Bihar, Chhattisgarh, Delhi, Haryana, Karnataka, Madhya Pradesh, Maharashtra, Odisha, Uttar Pradesh and Punjab. It is also recommended for release for *Kharif* in Chhattisgarh, Gujarat, Karnataka, Madhya Pradesh, Maharashtra, Odisha, Rajasthan and Tamil Nadu. Matures within 110-120 DAT. It has medium keeping quality and can be stored up to 3 months. Average marketable yield during Kharif season 18-20 t/ha and in *Rabi* is 26-30 t/ha.

Agrifound White

Selection made by National Horticultural Research and Development Foundation, Nashik. Bulbs are globular in shape, tight skin, silvery attractive white colour and 4-6cm in diameter. Total soluble solid more than 15 per cent and Pyruvic acid 12 micro mol/g. Mature in 110-130 days after transplanting and keeping quality good. Average yield 250-300 q/ha and suitable for *late Kharif* and *Rabi* season. Good variety for dehydration with drying ratio 7.10:1.

Agrifound Rose

Selection made by National Horticultural Research and Development Foundation, Nashik, from Chickballapur area of Karnataka. Variety is pickling type, grown in Kolar and Bangalore district of Karnataka and Cuddapah district of Andhra Pradesh exclusively for export. Bulbs are globular round in shape, deep scarlet red in colour and 2.5-3.5 cm in size. Mature in 95-110 days from seed sowing. Total soluble solid is 16-18 per cent and Pyruvic acid 10.27 micro mol/g. Average yield 190-200 q/ha. Variety is suitable for growing in *Kharif* in Cuddapah of Andhra Pradesh and in all three seasons in Karnataka.

Agrifound Red

Selection made by National Horticultural Research and Development Foundation, Nashik from Dindigul area of Tamil Nadu. Average size of cluster is 7.15 cm with weight of 65-67 g. Average numbers of bulblets per cluster 5.79. Size of bulblets is 3.64 cm, and average weight of single bulblets is 8.85 g. Colour of bulblets is light red. Total soluble solid is 17-19 per cent and Pyruvic acid 10.13 micro mol/g. Mature in 65-67 days after planting. Average yield is 180-200 q/ha.

Garlic

Agrifound White

Develop through mass selection by National Horticultural Research and Development Foundation, Karnal. Variety was notified by Government of India in 1989 vide notification number 280 (E) dated 13/04/1989. Bulbs are compact, silvery and white with creamy flesh. A bigger elongated clove with 20-25 in number and diameter of bulb is 3.5 to 4.5 cm. Variety is susceptible to purple blotch and stemphylium blight, which are common in the Northern parts of country, hence recommended for Maharashtra, and Madhya Pradesh where early crop escaping these diseases. Total soluble solid is 41 per cent, dry matter 42.78 per cent and pyruvic acid content is 27 micro moles per gram. Good for storage and average yield is 130 q/ha.

Yamuna Safed

Develop through mass selection by National Horticultural Research and Development Foundation, Karnal. This variety was notified by Government of India in the year 1991 vide notification number 527 dated 16/8/1991. Bulbs are compact, silvery white skin with creamy flesh, diameter 4.0-4.5 cm. Sickle shaped cloves with 25-30 in number, and diameter of cloves 0.8 to 1.00 cm. Variety is tolerant for purple blotch, stemphylium blight and thrips. Total soluble solid, dry matter and pyruvic acid content is 38 per cent, and 39.5 per cent and 29 micro moles per gram respectively. Good for storage and average yield 150-175 q/ha. Recommend to cultivation in all over the country.

Yamuna Safed-2

Develop through mass selection by National Horticultural Research and Development Foundation, Karnal. Variety was notified by Government of India in the year 1996 vide notification number 115 dated 10/2/1996. Bulbs are compact, attractive, white creamy flesh and diameter 3.5-4.0 cm. Number of cloves 35-40, diameter of cloves 0.75-1.4 cm and clove size index 1.75-2.5 cm^2. Total soluble solid, dry matter and pyruvic acid content is 38-40 per cent, 40-41 per cent and 26 micro moles per gram respectively. Average yield is 150-200 q/ha. Recommended for Northern India and gaining popularity in Punjab, Haryana and Madhya Pradesh.

Yamuna Safed-3

Develop through mass selection by National Horticultural Research and Development Foundation, Karnal. Variety was notified by Government of India in the year 1999, vide notification number 1052 (E) dated 26/10/1999. Leaves are wider than other varieties. Bulbs are creamy white and bigger sized and 4.5-6 cm in diameter. Diameter of clove 1.2-1.5 cm and cloves/bulb is 15-16. Total soluble solid, dry matter and pyruvic acid content is 38-42 per cent, 39-43 per cent and 25 micro moles per gram respectively. Medium in storage and average yield is 175-200 q/ha. Variety is suitable for export. The variety Gained Popularity in garlic growing areas of Madhya Pradesh, Maharashtra, Haryana, Gujarat, Punjab, Rajasthan, Uttar Pradesh and Chhattisgarh.

Yamuna Safed-4

Develop through mass selection by National Horticultural Research and Development Foundation, Karnal. This variety was notified by Government of India during 2006 vide notification number 597 dated 25/4/2006 for growing in Northern India. Plants are vigorous with wider green leaves. Bulbs are silvery white and bigger sized 3.5- 4.0 cm in diameter. Diameter of clove 1.2-1.25 cm, clove size index 2.50-3.00 cm^2 and number of cloves 20-25/bulb. Total soluble solid and dry matter content is 40-42 per cent and 41-42 per cent respectively. Average yield is 175-200 q/ha. Variety is recommended for growing in North India and gaining popularity for growing in Madhya Pradesh.

Yamuna Safed-5

Develop through mass selection by National Horticultural Research and Development Foundation, Karnal. Variety was notified by Government of India in 2012 vide notification number SO-2363 dated 04/10/2012. Dark green straight erect leaves, sturdy plant growth, erect shape-wide 8-10 leaves/plant. Bulbs compact, attractive creamy white colour, creamy flesh, big size bulbs (4.5-5.0 cm in diameter) and cloves, oblong shape, number of cloves per bulbs 22-30. Variety is table purpose, pungent with good aroma. Total soluble solids and dry matter content is 39 per cent and 41 per cent respectively. Good storer and better adaptability and tolerant to stemphylium blight, purple blotch and thrips under field condition. Average yield is 150-160 q/ha. Suitable for processing purpose and recommended for cultivation in Sikkim, Meghalaya, Manipur, Nagaland, Mizoram, Tripura, Arunachal Pradesh, Andaman and Nicobar Island, Punjab Tarai region of Uttar Pradesh, Uttarakhand, Bihar, Jharkhand, Gujarat, Rajasthan, Haryana and Delhi.

Yamuna Safed-8

Garlic variety, **Yamuna Safed 8 (G-384),** released and notified through Crop Standard Notification and Release of Varieties of Horticulture Crops, Ministry of Agriculture Cooperation and Farmer welfare, Government of India vide **Notification Number S. O. 268 (E) 28th January 2015** for **Zone II** (Jammu and Kashmir, Punjab, Delhi, Haryana and Rajashthan). Dark green plant, plant straight, erects leaves, Average bulb diameter 4.30 to 5.00 cm. Mature in 150-160 days. Average yield potential is 175-200 q/h and moderate storability.

Yamuna Safed-9

Garlic variety, **Yamuna Safed 9 (G-386),** released and notified through Crop

Standard Notification and Release of Varieties of Horticulture Crops, Ministry of Agriculture Cooperation and Farmer welfare, Government of India vide **Notification** number SO-3666 (E) dated 06/12/2016 for **Zone II** (Jammu and Kashmir, Punjab, Delhi, Haryana and Rajasthan). Dark green plant, plant straight, erects leaves; Average bulb diameter 4.80 to 5.50 cm. Average numbers of cloves per 22-30 and clove diameter is 1.30-1.50 cm. Mature in 150-160 days. Average yield potential is 200-250 q/h and moderate storability.

Agrifound Parvati

Garlic variety, **Agrifound Parvati (G-313)** Released and Notified notified through Crop Standard Notification and Release of Varieties of Horticulture Crops, Ministry of Agriculture Cooperation and Farmer welfare, Government of India vide **Notification Number S. O. 268 (E) 28th January 2015** for **Zone-I** (Jammu and Kashmir, Uttarakhand, Himachal Pradesh and Tamilnadu. Variety is long day type and suitable for cultivation only in mid and high hill of Northern states. Bulbs are of bigger in size 5.0-6.5 cm in diameter, creamy white colour with pinkish tinge, diameter of cloves 1.5-1.8 cm, 10-16 cloves/bulb. Total soluble solid, dry matter and pyruvic acid content is 35-36 per cent, 37-38 per cent and 22-23 micro moles per gram. Medium in storage and average yield is 175-225 q/ha.

Agrifound Parvati-2

First garlic variety, **Agrifound Parvati 2 (G-408),** released and notified through Crop Standard Notification and Release of Varieties of Horticulture Crops, Ministry of Agriculture Cooperation and Farmer welfare, Government of India by vide **Notification Number S. O. 268 (E) 28th January 2015** for cultivation in **Zone-I** (Jammu and Kashmir, Uttarakhand, Himachal Pradesh and Tamilnadu). Variety is long day type and suitable for cultivation in Himalayan hills of Himachal Pradesh, Uttarakhand, Jammu and Kashmir and high altitude of North eastern states of Sikkim *etc.* Bulbs are of bigger in size 5.0-6.0 cm in diameter, creamy white colour, diameter of cloves 1.5-1.6 cm, 1-14 cloves/bulb. Total soluble solid, dry matter and pyruvic acid content is 36-37 per cent, 37-38 per cent and 22-23 micro moles per gram. Average yield is 175-200 q/ha.

Scientific Cultivation for Enhance the Profitability to the Farmers

Onion production depends upon the principles of biotic, abiotic and socio-economic factors constraining the realization of potential yield. Apart from various climatic conditions, these factors vary from region to region and season to season. Major factors playing an important role in the production of onion are given below:

Climate and Soil

Onion is grown under a wide range of climatic condition, *i.e.* temperate, tropical and subtropical. It grows well under mild climate without extreme heat or cold or excessive rainfall. Its plant growth, bulb development and bolting or flowerings are influenced by photo period (day length) and temperature for good vegetative growth.

It does not thrive when average rainfall exceeds 75–100 mm during monsoon period. The ideal temperature for its vegetative growth is 12.8–23.0°C. For bulb development, it requires long days and still higher temperature *i.e.* 20°C–25°C. Warm temperature favours good development of bulbs. In India, onion varieties grown in plains are short day types, requiring 10–11.5 hours' day length; while long day types require 13–14 hours' day length are cultivated in hills. In *Rabi* season a sudden rise in temperature may result in early bulb maturity and smaller sized bulbs. The long day varieties do not bulb under short day conditions, whereas short day varieties if planted under long day condition develop bulb. As temperatures conditions are never exactly alike year after year in the same place or field onion varieties cannot be expected to perform constantly even though practices are similar. Temperature is more important than day length in seed production, while photoperiod is more important than temperature in bulb formation.

Soil for onion cultivation should be rich in humus with good drainage. It can be well grown on light loam, sandy loam or clay loam, deep friable and fertile soil rich in organic matter. The soils which have capacity to retain enough moisture and at the same time be favorable enough to be easily cultivated and to help for proper development of bulbs are most desirable soil for onion cultivation. The optimum soil pH is 5.8-7.5. Highly alkaline or saline soil is not suitable for onion cultivation. The soil having higher water table is also not suitable for onion cultivation.

Cropping System

The onion is crop generally taken by growers after the previously heavily manured grown crop. The *kharif* or *late kharif* crops are being grown after coriander, French bean, sorghum, pearl millet, chilli and groundnut. *Rabi* crop is being taken after groundnut, cowpea, potato, *etc.* Onion also being grown as an intercrop in sugarcane, chilli, cotton in some pocket of Haryana, Maharashtra, Uttar Pradesh and Karnataka.

Methods of Planting/Propagation

Raising Seedlings

Onion seed is generally sown in raised nursery bed (15–22 cm height) for transplanting in the field. The width of nursery bed should be 0.65 m and length 3–4 m. The distance in between two beds should be kept about 45–60 cm. The surface of beds should be smooth and well leveled. The best time for sowing the seed for *kharif* is May-June, August-September for *late kharif* and October-November for *rabi* in the plain all over the country. Sowing of onion seeds should be done in lines at a proper spacing (5-7 cm row-to-row). Before sowing, seed should be treated with thiram @ 2.0 g/kg to avoid damage of seedlings from damping off disease. The soil of nursery should also be treated with thiram or captan @ 4.0-5.0 g/m^2 area.

The nursery soil should be irrigated 20-25 days, before sowing and covered with 250-gauge transparent polythene for soil solarization. The application of *Trichoderma viride* fungal culture @ 1250 g/ha is also recommended to manage damping off and raising the healthy nursery. The *Trichderma viride* should be mixed in 25 times decomposed sieved FYM or sieved moist soils for application in field. The seeds after sowing should be covered with fine powdered FYM or compost followed by light watering by water can. After that beds, should be covered with dry straw or grass or sugarcane leaves to maintain required temperature and moisture. The watering should be done by water can as per the need till germination is completed. The mulching materials should be removed immediately after germination. Any delay in removal of mulch, results in development of lanky seedlings. The seeds of onion Agrifound Dark Red coated with DAP + $ZnSo_4$ + Bavistin as well as with DAP + Borax + *Trichoderma* and sown in July and harvested in November gives highest gross yield.

Weeding and hoeing are done to manage weeds timely. Stomp (Pendimethalin 30 EC) @ 2.5 litre/ha or Goal (Oxyfluorfen 23.5 per cent EC) @ 600 ml/ha can be applied even 15-20 days before sowing to manage the weeds economically during rainy season wherever there is labour scarcity. The seedlings become ready for transplanting after 6-7 weeks for *kharif*, and 8-9 weeks old seedlings for *late kharif* and *Rabi* seasons.

Raising Small Bulblets and Planting

The bulblets raised from *kharif* onion varieties like Agrifound Dark Red, Baswant-780, N-53 and Arka Kalyan in the previous season are used for planting. Seeds @ 10-12 g/m2 are sown on raised beds or in flat beds depending upon the soil like raising of nursery meant for transplanting method. The best time of sowing of seeds for getting quality bulblets is mid-January to early February depending upon the area. All others operations for raising of bulblets are the same as that of nursery for transplanting. The plants are allowed in nursery bed up to April-May till there is top fall. Harvesting is done along with the tops and selected bulblets (1.5-2.5 cm) are stored by hanging method till August in a well-ventilated room/go down. Bulblets below 1.5 cm if planted do not establish and sprout but rot. Large sized bulblets result in more doubles and bolters and increased cost of production also.

Seed Rate, Manuring and Fertilization

About 7-8 kg seed for common big onion is sufficient for one hectare. The soil for onion growing should be liberally manured and fertilized. Application of 20-25 tonnes of FYM/ha in soil is considered adequate. The requirement of nutrients depends on soil type, region of growing, varieties and removal of major nutrients. Farmyard manure should be applied one month before transplanting/planting or sowing and mixed well in soil. 100:50:50:40 (NPKS) recommended for good quality onion production. Whole quantity of phosphorus, potash and Sulphur and half of nitrogenous fertilizers should be mixed in soil before transplanting. Rest half doses of nitrogenous fertilizers should be given as topdressing in two equal split doses. The first dose should be applied 30 days after transplanting, whereas second dose 45 days after transplanting. The topdressing must be completed before initiation of bulbing. Delayed application results in thick neck or doubles. At Karnal, 50 per cent nitrogen at transplanting, 25 per cent after 30 days and 25 per cent after 60 days of transplanting gives higher yield in *rabi* onion.

In addition, to fertilizer, application of micro-nutrients is beneficial in improving the quality of bulbs. Application of zinc @ 1.2 and 3.0 ppm is increases yield with enhanced quality of bulbs, increasing total soluble solids, sugar and ascorbic acid. Copper and boron @ 13.4 kg and 1.8 kg/ha, respectively are beneficial in increasing yield. At Nasik, application of Zn @ 3.0 ppm, copper @ 1.0 ppm and boron @ 0. 5 ppm 45 and 60 days after planting in *kharif* onion crop gives higher return. The application of poly feed (19:19:19:TE) @ 1 per cent at 15,30 and 45 days as foliar sprays and multi-K (13:0:46) @ 1.0 per cent at 60,75, and 90 days improves yield as well as storability in onion. Foliar spraying of NPK (19:19:19) 15, 30 and 45 days after planting (DAP) and NPK 13:0:45, 60, 75 and 90 days after planting in onion bulb crop are recommended for getting higher yield with improved quality of bulbs.

Land Preparation and Transplanting

The field is ploughed to a fine tilth by giving four ploughings with a sufficient interval between two ploughings. The ploughings should be done by tractor drawn implements or *deshi* plough. It should be shallow as most of the onion root penetrates to a depth of not more than 5-10 cm. The planking should be done for proper levelling. The field is then divided into beds and channels. The normal width of a bed should be about 1.80 m. The length may vary as per level of land but it is better to have small sized beds for proper irrigation and other cultural practices. If beds are wider than 1.8 m, enter in bed for weeding and hoeing as also topdressing or plant protection resulting in injury to plants. The width of beds should be such that one can arrange intercultural operations by sitting on the bunds. If length is more, levelling becomes difficult. Unlevelled land results in water logging at one point and no water at another point. The water logging affects production and quality of bulbs both adversely. In *kharif* season, planting should be done on raised beds to avoid water logging.

Seedlings are usually transplanted in flat beds. Transplanting on raised beds or on both sides of ridges is, however, better for Kharif or rainy season crops. Flat beds of 1.8 m width and 7.2 m length depending upon level of land, soil type and irrigation method, are prepared. About 6-7 weeks old seedlings are transplanted in *kharif* season, whereas 8-9 weeks old seedlings are ideal for *Rabi* season. The bolting increases in overage planted seedlings and establishment of seedlings become poor in younger seedlings.

The best time of transplanting for *kharif* onion is July-August in Maharashtra and August in northern part of country. The best time for *rabi* onion transplanting, in Northern and Eastern India is from December end to first week of January. Mid-December is best time for other parts of country. In Nashik and Karnal areas mid-December planting of Agrifound Light Red and Pusa Red onions gives higher yield. Dipping of seedlings root in fungicides (carbendazim @ 0.1 per cent) and insecticide (monocrotophos @ 0.1 per cent) solution before transplanting helps in better establishment of seedlings.

Spacing and Irrigation

The spacing depends upon variety and size of bulbs. In common big sized onion 15.0 cm from line-to-line and 10.0 cm from plant-to-plant is considered is best. A spacing of for Karnal like areas, 150 kg N and 15.0 cm x 10.0 cm spacing and fro Nashik areas 100 kg N and 15.0 cm x 10.0 cm spacing are ideal. For raised bed planting method, a spacing of 10.0 cm x 10.0 cm to 12.0 cm x 10.0 cm is more suitable for higher yield and storability of bulbs.

The cylindrical leaf of onion is effective for dissipating heat in air since it has highest heat transfer per unit surface area. Stomata close at low value of potential evapo-transpiration compared with other species. Thus narrow, vertically oriented and cylindrical leaves are well adopted to dissipate energy as heat without any excessive rise in leaf temperature.

The irrigations depend on several factors such as crop growth, soil type and planting season. Onion is shallow rooted crop. Its roots system is normally restricted to top 3.0 cm and roots penetrate seldom deeper than 15.0 cm. The water requirement of the crop in the initial growth period is lesser. The irrigation interval of 15 days in *rabi* season up to 60 days and 8 days afterwards gives more yield.

First irrigation is necessary immediately after transplanting, particularly in rainy season in Northern India where temperature at the time of transplanting is very high. If irrigation is delayed due to some reasons there is 80-90 per cent mortality. Wherever, there is erratic power supply in rainy season, transplanting is recommended 8 hours after irrigation, particularly in Karnal like areas. In *kharif* crop season depending upon the rains 6-8 irrigation are enough, 8-10 irrigations for *late kharif* and 10-12 irrigations for *Rabi* season crop required. At bulb formation, irrigation is necessary and moisture stress at this stage results in low yield.

Saturating the soil surface always favours the development of purple blotch disease in onion, particularly at full vegetative growth. The field should not be kept dry for a long period which otherwise results in splitting and force the maturity of bulbs. Frequent and light irrigation at weekly intervals in general (2.0-acre inch per irrigation) results in good bulb development, increasing yield also. In *kharif*, irrigation at 2-week interval and N @ 200 kg/ha gives highest yield of quality bulbs. Heavy irrigation by flooding method gives poorly developed bulbs.

Sprinkler and drip irrigation methods give significant increase in yield. Drip irrigation with raised bed planting is recommended for getting higher yield with improved bulb quality in Maharashtra. Irrigation at 1.25 ID/CEP ratios (10 irrigations) in *rabi* season is recommended for Karnal areas. At MPKV Rahuri, sprinkler irrigation at 75 cm CPE at 40 mm depth and in drip irrigation in *rabi* season 21 cm water is beneficial for higher yield over surface irrigation system. The monitoring of water through drip system is essential to keep optimum subsoil and moisture in beds for better bulb development. Irrigation should be stopped when tops mature and start falling in *rabi* season. In *kharif* season it should be stopped to 10 days before harvesting. Last irrigation, 10-15 days before harvesting gives lower storage losses.

Weed Management

Weeds are to be kept down to get a good yield of onion. The onion weeds common in *kharif* season are *Cyperus rotundus, Cynodan dactylon, Chaenopodiumn album, Amaranthus sp. Scoparia sp, Convolvulus sp Echinocholoa colonum etc.*, while *Rabi,* in addition to these, *Argimone maxicana, Avena sativa Allium sp.* (*wild onion*) and *Portulaca sp.* Also, causes losses to onion. Onion being closely planted and a shallow-rooted crop hand weeding, particularly when crops is in full vegetative stage, is difficult and sometimes because of shortage of labour and it is expensive. It is difficult to manage weeds manually. Therefore, it is recommended to use weedicide along with one hand weeding at critical stage. Use of stomp @ 3.5 litre/ha applied 3 days after transplanting and one hand weeding at 45 days in *rabi* season control weeds effectively, while in *kharif* onion, stomp @ 3.5 litre/ha with one hand weeding is found effective at Nashik. Oxyfluorfen @ 150-250 g a. i./ha with one hand-weeding is also found effective.

Growth Regulators

Spraying of cytozyme @ 0.2 per cent 15 and 45 days after transplanting at Nashik is effective. Dipping of seedling roots for 20 minutes before transplanting and foliar spraying at Karnal are found more effective in bulb development and higher yield of *kharif* onion. The spraying of MH @ 2000 ppm to check premature bolting and sprouting in storage is recommended 75 days after planting for *kharif* onion at Nashik and 3000 ppm at Karnal. Spraying of MH @ 2500 ppm just before neck fall in *Rabi* onion improves the storage quality of bulbs. Use of growth retardants like cycocel, as well as ethrel and GA is also recommended for increasing yield but its response varies for variety to variety. The application of cycocel @ 2500 ppm after 75-80 days, if foliage growth is more can be given to attain bulb development by stopping vegetative growth. However, use of growth regulators to enhance bulb development under adverse weather condition may be promising under certain conditions.

Harvesting and Yield

Onion is harvested depending upon the purpose for which its crop is planted. Harvesting takes 45-90 days from field setting for green onions and 65-150 days for bulb depending upon the varieties. Bulbs are mature when the neck tissues begin to soften and tops are about to abscise and decolorize. Development of red pigment and characteristic pungency of variety are also important harvest indices of onion.

Onions for use in green stage are harvested as soon as they reach edible size. The plants are pulled by hand, the roots are trimmed and the outside skin is peeled off, leaving the leaves clean and lower portion above the bulb white. The onions are washed, sorted and tied in bunches. The size of bunch depends on local market preference. If onions are to be transported to long distant market, these are packed in crates with crushed ice to avoid discoloration and wilting of leaves at high temperature.

Onions for sale as dried bulbs or for storage should be harvested after tops have started falling over. Best time to harvest *Rabi* onion is one week after 50 per cent tops have fallen over. In *Kharif* season, since tops do not fall, soon after the colour of leaves changes to slightly yellow and tops start drying, red pigmentation on bulbs develop and true shape and size develop, bulbs are harvested and kept in windrow for drying the tops. Small picking and multiplier onions are harvested when 50 per cent -75 per cent of tops have fallen over. The leaves are cut leaving 2.0-2.5 cm tops above the bulbs after complete drying. Leave the tops intact until complete drying increases dry-matter content which might be due to more loss of water from the bulbs with intact foliage or to movement for materials from the tops to bulbs. If tops are cut too close, the neck does not close well and decay organisms have easy access to bulbs. Early harvesting is results in sprouting of bulbs, late harvesting results in formation of secondary root during storage. In *Kharif* season, late harvesting results in doubles and bolting. Common big onion gives about 25-45tones/ha yield.

Brief Garlic Production Technology

Garlic successfully grown in fertile, well drained loamy soils with pH range is 6.0–7.0 and plant by four methods *i.e.* Dibbling, Furrow, Broad casting and Seed drill. The suitable time of planting is October-November. The late planting may affect physiological maturity resulting in loss of yield and early planting also re-sprouting if winter is prolonged thereby affecting the bulb quality.

About 5-7 q cloves of 8-10 mm diameter are required for one hectare planting and for good yield, select bigger cloves preferably from the outer side of the bulbs. Plant nutrients apply based on the soil test, but generally recommended to apply 25 t/ha well decomposed FYM at the time of field preparation and mix well in the soil. Apply 100 kg nitrogen, 50 kg phosphorus, 50 kg potassium and 40 kg sulphur/ ha through available source of chemical fertilizers. The whole dose of phosphorus, potassium, sulphur and half dose of nitrogen should be applied before planting and rest half dose of nitrogen should be divided in two parts and applied at 30 days and 45 days after planting. Planting distance 10 x 7.5 cm spacing has been recommended and irrigation at an interval of 8-10 days during vegetative growth and 10-15 days during maturation. The crop is harvest when the tops turn yellowish or brownish colour and show signs of drying up and bend over. Harvesting at 100 per cent neck fall gave minimum storage losses and field curing for 3-5 days by windrow method was found to be beneficial. Bulbs are taken out along with the leaves and kept in curing about 15-20 days under field condition. After that the necks are cut by leaving 2.5-3.0 cm from bulbs and sent in the market after proper sorting and grading for sale. It can be stored in bundles along with foliage or in hessian bag or leno bags after foliage cut in damp proof ventilated godowns up-to six months. Garlic is affected by Purple Blotch: On leaves, as small whitish sunken lesions with purple colour centers that rapidly enlarge. Mancozeb @ 2.5 g/lit sprayed at regular intervals of 15 days after disease appearance. Stemphylium Blight: Infection appears as small, yellow to orange flecks or streaks on the leaf. They usually turn gray at the center, brown to dark olive brown. Mancozeb @ 0.25 per cent or chlorothalonil (Kavach) @ 0.2 per cent along with deltamethrin (Decis) @ 0.10 per cent and sticker triton at fortnightly interval. Thrips: Very minute yellow to dark brown in colour and life of insect is 8-10 days. It is found in the axil of green leaves where it sucks juice of early emerging leaf. The infected plants show stunted growth with twisted leaves. Malathion @ 0.1 per cent or metasystox @ 0.1 per cent and cypermethrin @ 0.01 per cent or deltamethrin 2.8 ai @ 20 ml/ha should be sprayed along with triton or sandovit.

Chapter 15

Doubling Farmer's Income through Food-Fodder Based Cropping Systems in Rainfed Areas

Sunil Kumar

HOD, Crop Production Division
ICAR-Indian Grassland and Fodder Research Institute,
Jhansi – 284 003, U.P.

Introduction

Agriculture is the main stay of Indian economy and gross capital formation in agriculture and allied sector as per cent to gross value added (GVA) was 15.8 per cent in 2015-16 (at current 2011-12 prices) (Economics Survey, 2015-16). The average monthly income per capita from farming increased from Rs. 1,060 in 2003 to Rs. 3,844 in 2013, according to the report Situational Assessment of Agricultural Households by the NSSO, a compounded annual income growth rate of 13.7 per cent. To double the income of farmers *i.e.* Rs. 7,688 by 2022, in nominal (numerical) terms—which do not take inflation into account—would require a 15 per cent compounded income growth rate, which is a marginal increase over the achieved increase from 2003 to 2013. However to increase the income in real terms would imply restructuring agriculture processes and policy interventions.

Livestock sector has a share of about 27 per cent to total GDP contributed by Indian agriculture. According to 2012 livestock census, India has world largest livestock population numbering 512 million. With this livestock population, Indian ranked first in milk production with a production of 155 mt during 2016-17. Productivity of Indian livestock is much lower compared to developed countries. The main causes behind lower output are under-nourishment of livestock and lack of availability of good quality of green and dry fodder and concentrates. According to IGFRI Vision 2050, the demand will reach to 1012 million tonnes of green fodder and 631 million tonnes of dry forage by the year 2050. At the current level of growth in forage resources, there will be 18.4 per cent deficit in green fodder and 13.2 per cent deficit in dry fodder in the year 2050. The availability of green fodder will be more difficult under changing climatic conditions prevailing in the country in terms of floods, drought, cyclones *etc.* will lead to livestock rearing will further become costlier.

Table 15.1: Demand and Supply Estimates of Green and Dry Forages (million tonnes)

Year	*Demand*		*Supply*		*Per cent Deficit*	
	Green	*Dry*	*Green*	*Dry*	*Green*	*Dry*
2010	816.8	508.9	525.5	453.2	35.66	10.95
2020	851.3	530.5	590.4	467.6	30.65	11.85
2030	911.6	568.1	687.4	500.0	24.59	11.98
2040	954.8	594.9	761.7	524.4	20.22	11.86
2050	1012.7	631.0	826.0	547.7	18.43	13.20

Source: IGFRI Vision (2050)

Concept and Overall Strategy for DFI by 2022

The Hon'ble Prime Minister while addressing a farmers' rally in Bareilly, Uttar Pradesh, on 28th February, 2016 shared his vision on doubling the farmers' income by the time the nation celebrates her 75 years of independence in the year 2022. He subsequently reiterated this by shaping his vision around the following framework:

1. Special focus on irrigation with sufficient budget, with the aim of "***Per Drop More Crop***".
2. Provision of quality seeds and nutrients based on soil health of each field.
3. Large investment in Warehousing and Cold Chains to prevent post-harvest crop losses.
4. Promotion of value addition through food processing.
5. Creation of National Farm Market, removing distortions and e-platform across 585 Stations.
6. Introduction of a New Crop Insurance Scheme to mitigate risks at affordable cost.
7. Promotion of ancillary activities like poultry, beekeeping, and fisheries.

The broad approach towards achieving the vision of "***Doubling of Farmers' Income* (DFI)**" by the year 2022 would be:

- ☆ Increasing per unit productivity
- ☆ Reducing cost of cultivation
- ☆ Increasing net-farm income from each unit
- ☆ Ensuring higher market return of the farmers' produce

Different Approaches for Enhancing Forage Production

The productivity is a consequence of the interactions between genotypes and environment. The phenomenal increase in productivity is possible mainly through better seeds, fertilizers and better agronomic technology (Kumar *et al.*, 2013). The varieties with early bulkiness and growth rhythm matching physiology with concurrent weather variability for higher rate of dry matter accumulation and longer leaf area duration in relation to climatic elements will find the place in the system. Quick regeneration potential, flexibility to defoliation management and greater persistency under adverse climatic conditions would be an added advantage of the forage crops. In crop management arena timely sowing is very important in forage crops as these have very limited growth period. The seed rate and spacing may need adjustment depending upon adverse weather situations. The moisture has been a limiting factor; hence, judicious use of water and its management will remain to hold prime importance to optimize the productivity and resource use in the years to come. Irrigation scheduling may be a pertinent approach to water saving because it takes care of weather conditions for efficient use by the crops. For controlling insect-pests and diseases, it would be better to take precautionary measures based on predicted weather based advisory services. In longer term, anticipatory adaptation might include the development and use of new crop varieties that offer advantages for the anticipated future climate variations, or investments in new water- management and irrigation infrastructure as insurance against potentially less reliable rainfall.

Management Strategies

In order to sustain the profitability of livestock, the availability of quality feed and forage is of paramount importance. The flexibility of agricultural systems suggests that, there is a significant inherited potential to adapt to climate change. Appropriate strategies can minimize the yield losses that might result from climate change, or improve yields where climate change is supportive. But developing adaptation strategies exclusively for minimising the negative impact of climate change may be risky considering the number of uncertainties associated with its spatial and temporal magnitude. Therefore, these adaptations can be at the level of individual farmer, farm, society, village, watershed or even at a national level. Farm level adaptation options can be made in tillage practices, planting and harvest dates, crop rotations, selection of suitable crops and crop varieties for cultivation, water consumption for irrigation, fertilizer management and use of biofertilizers can reduce the impacts of climate change on forage crops. Improved crop management, as well as, better risk management through an early warning system and crop insurance policies can also be beneficial measures to reduce the vulnerability of the farmers.

Forage Production from Rainfed Lands

Under changing climatic situations selection of production systems should be done on the basis of rainfall pattern, available irrigation facilities and most suitable forage species and cropping system which support under aberrant weather situation. Such type of production systems can be grouped in to following sub headings:

The selection of forage crops and their varieties *viz.*, pearl millet, sorghum, cowpea, cluster bean, ricebean, TSH, guinea grass and pearl millet + cluster bean, sorghum + cowpea *etc.* in cropping system may prove most productive under rainfed climatic situations.

Model-I

Perennial based fodder - food production System: Subabul + Trispecific hybrid - sorghum (fodder) + pigeonpea (grain)

i. Production potential: 53.3 t/ha green fodder grain -0.4 t/ha, sticks – 0.8 t/ha
ii. Cereal l: legume : 80 : 20
iii. Adoption area: Semi arid region
iv. Clientele group: Small and medium farmers
v. Water requirement: areas up to 500 mm rainfall
vi. Livestock's support: 2-3 milch animals

Model-II

Seasonal based fodder - food production System: Sorghum (Grain) + cowpea (Fodder)

i. Production potential: Grain- 2.5 t/ha, Green fodder – 17 t/ha
ii. Cereal : legume: 67 : 33
iii. Adoption area: Semi arid region
iv. Clientele group: Small and medium farmers
v. Water requirement: areas up to 500 mm rainfall
vi. Livestock support: 2-3 milch animals

These systems have green fodder production potential of 53.3 t/ha and 13.28t dry fodder besides producing 0.4 t/ha pigeon pea grain and 0.8 t/ha dry sticks having fuel value. Technology can sustain 2-3 ACUs with quality fodder. It also provides nutrient rich pulse for human consumption and fuel wood. It is also capable of prolonging the fodder availability up to May (dry period).

Integrated Farming Systems Models

Model – 1: Rainfed Crop Dominated IFS

Enterprise/Activity	*Area (ha)*	*Crop (ha)*	
A. Crops	0.5	*Kharif*	*Rabi*
		Groundnut, 0.3	Barley, 0.2
		Sorghum or Maize (cob), 0.2	Gram, 0.2
			Oat, 0.1 (F)
B. Agri-horticulture	0.2	Ber + Sesame-gram	
C. Silvipasture	0.2	Subabool+Cenchrus/TSH + S. Seabrana (strip/paired plantation)	

Enterprise/Activity	Area (ha)	Crop (ha)
D. Animals	-	5 goats with kids up to 6 months
E. Farm pond	0.1	For rain water harvesting and recycling
Total	1.0	-
Control	0.25ha (50x50m)	Prominent cropping practice of area i. Blackgram+sesame-fallow ii. Fallow-gram+mustard
Boundary plantation of ber+subabul+caronda and bund plantation of napier/guinea/TSH * If rainfall delayed, groundnut to be replaced by blackgram/sesame		

Model – 2: Rainfed Livestock Dominated IFS

Enterprise/Activity	Area (ha)	Crop (ha)	
		Kharif	Rabi
A. Crops	0.4	Blackgram, 0.2	Barley, 0.2
		Sorghum or Maize (cob), 0.2	Oat, 0.2 (F)
B. Agri-horti-pasture	0.2	Ber + TSH+Stylo+Bajra-gram (Strip plantation)	
C. Agri-silvi-pasture	0.3	Subabool+guinea/TSH + S. Seabrana+Bajra-mustard (strip/paired plantation)	
D. Animals	-	1C + 2 goats with kids up to 6 months or 8 goats + kids	
E. Farm pond	0.1	For rain water harvesting and recycling	
Total	1.0	-	
Control	0.25ha (50x50m)	Prominent cropping practice of area i. Blackgram+sesame-fallow ii. Fallow-gram+mustard	
Boundary plantation of ber+subabul+caronda and bund plantation of napier/guinea/TSH * If rainfall delayed, groundnut to be replaced by blackgram/sesame			

Economic output: Generally rainfed farmers are getting Rs. 20000 to 25000/ ha under normal conditions which has enhanced to Rs. 40000 to 50000 through this model.

Recycling: Rainwater, FYM, farm waste composting, *etc.*

Resilience: Rainwater harvesting and recycling, introduction of perennial components (Subabul, perennial TSH, Stylo, Ber) and animals.

Forage Production from Non-arable Lands

Under climatic change, it is predicted that the length of the growing and grazing season will increase with the extensions occurring at both ends of the season. Production in terms of quality and quantity of natural grassland can be improved with the introduction of range legumes species. These legumes are *Stylosanthus hamata, S. seabrana, Atylosia scarabaeoids, Clitoria ternatea, Desmodium tortosum, Glycine javanica, Lablab purpureus, Macroptilium atropurpureum, M. lathyroides, Mimosa invisa, Stizolobium deringianum, Stylosanthes humilis, S. guianensis, Sesbania sesban* and *Vigna luteola* which are generally introduced in natural grassland to enhance the forage production. The local grasses, although hardy but they have low yield potential, low protein content and are less palatable. Most of the grassland situated in non-cultivable dry region is very low in legume component. The introduction of legume is very important for improvement in production potential in term of quantity and quality.

Grass species belong to C_3, C_4 and CAM types; maximum grasses are C_4 type, and C_4 type grasses are widely adapted to drought, salinity, water logging and climate changes than C_3 grasses. Elevation of CO_2 concentration affect the higher biomass which is pre requisite of forage crop species. C_3 grasses are more benefitted by increase CO_2 concentration than C_4 grasses; C_4 grasses are more suitable in cooler regions than C_3 grasses due to rising temperature. Therefore, adaptation of forage species particularly perennial species has ability to produce more forage yield under changing climate and reduces the ill effect of climate change through carbon sequestration (Kaul *et al.*, 2010). The other benefit of these perennial forage crops is that, they are established and grown easily under rainfed conditions with less agricultural practices (one-time establishment). Dry area farming has well proven as option for climate change to mitigation and adaptation. It also reduces the interculturing operation in field, which has been recognized the main source of lease of CO_2.

Fodder from Alternate Land Use Systems

The Millennium Ecosystem Assessment has categorized the agroforestry ecosystem services into provisioning services (fuel-wood, fodder, timber, poles *etc.*), regulating services (hydrological benefits, micro-climatic modifications), supporting services (nutrient cycling, agro-biodiversity conservation), and cultural services (recreation, aesthetics). Agroforestry land use increases livelihood security of rural masses and reduces vulnerability to climate and environmental change. There are ample evidences to show that the overall (biomass) productivity, soil fertility improvement, soil conservation, nutrient cycling, microclimate improvement and carbon sequestration potential of an agroforestry system is generally greater than that of an annual system (Pathak and Sunil Kumar, 2004).

Forages on Bunds

If 10 per cent Farm Bunds of the country utilized for fodder production then 17.8 mt green fodder production per year can be obtained.

- ☆ Suitable grasses- Perennial grasses *viz.*, NB hybrid, TSH, Setaria and Guinea grass.
- ☆ 1400 to 2350 kg green fodder/year over single grass row of one-hectare area
- ☆ Fodder yield recorded from different farm demonstrations indicated 1.75 to 2.50 kg green fodder per meter per cut and on an average in 4 cuttings 7.0- 11.0 q green fodder per 100 metre bund length is possible in a year.

Besides additional farm productivity, it also works as a guard crop for main crop, reduces runoff loss of water and controls soil erosion.

Non-conventional Fodders

Fodder beet, lathyrus, vicia, forage cane and spineless cactus may be popularized under different niches as an alternate source of fodder in addition to available cultivated fodder species. On the basis of energy, forage beet can give highest quantity of yield compared to rest of the fodders. Spineless cactus may be planted on boundaries of fields and provide green fodder even under harsh conditions.

Spineless Cactus: A Non-conventional Fodder Resource for Livestock

The spineless cactus is promoted as non-conventional fodder during lean periods particularly 3-4 months of summer in degraded lands, farm boundaries and pasture lands through community involvement in Bundelkhand region. The cactus accessions obtained from ICARDA (Mexico, Brazil and Morocco) were evaluated and established in semi-arid region of Bundelkhand. The cactus contains 7-9 per cent CP and 85-92 per cent water which is a potential fodder resource during summers, drought or during winter lean period. It has potential of producing 80 kg of fresh weight per plant (3-year plantation) and 100m row length planted 1 m spacing of cactus can provide 4 months' fodder for 1 ACU. Cactus can be successfully fed to livestock as supplementary feed (30 per cent Replacement) along with green fodder, hay and straw, but not as sole feed. It can successfully be grown in degraded lands, farm and field boundaries. It provides green fodder as supplement to livestock during lean periods (winters, harsh summers or during dry spells/droughts) of semi-arid region.

Key Issues

- Augmenting production by improved climate resilient forage production technologies strategies.
- The watershed management programme for efficient use of water and natural resources under changing rainfall pattern.
- Increasing the income from agricultural enterprises by suitable actions such as accelerated development of location-specific fertilizer management practices, improved fertilizer supply and distribution system, improved water and fertiliser use.
- Prediction based risk management through an early warning system and policies that encourage crop insurance.
- Use of information technology for dissemination of the forage production technologies.
- Development and adoption of an early warning system for insect-pest and disease management.
- Facilitate greater adoption of scientific and economic pricing policies.

- ✰ Consider financial incentives and package for improved land management.
- ✰ Collaborative management of natural resources by including different stakeholders such as farmers, pastoralists, herders and policy makers *etc.*
- ✰ Community involvement in adaptation strategies.
- ✰ Development of degraded waste lands through aforestation, reforestation, restoration for fodder production.
- ✰ Use of ITKs (Traditional Indigenous knowledge) of local communities and indigenous peoples for control of natural calamities and conservation for future geenration.

References

Economic Survey 2015-16. Department of Revenue, Ministry of Finance, Government of India, New Delhi.

IGFRI Vision 2050. ICAR-Indian Grassland and Fodder Research Institute, Jhansi (UP), India pp. 1–25.

Kaul, M., Mohren, G. and Dadhwal, V. 2010. Carbon storage and sequestration potential of selected tree species in India. *Mitigation and Adaptation Strategies for Global Change* 15(5): 489- 510.

Kumar Sunil, Faruqui SA and Singh GK 2013. Energy use analysis of food-forage based cropping system in semi arid environment of India, *Indian Journal of Agricultural Sciences* 83(11): 1236-1238.

Pathak, P.S. and Sunil Kumar. 2004. Forage and grazing resources in different agro-climatic regions in India In : S.S. Kundu, A.K. Mishra and P.S. Pathak (eds.). *Buffalo production under different climatic regions.* International Book Distribution Co. Lucknow p. 1-41.

Singh, Gurbachan, 1995. An agroforestry practice for the development of salt lands using *Prosopis juliflora* and *Leptochloa fusca*. *Agroforestry Systems* 27: 61-75.

Chapter 16

Safe Pest Management Approaches in Production of Vegetables

Rakesh Pandey and M.K. Mishra

Department of Entomology,
Banda University of Agriculture and Technology, Banda, U.P.

The vegetable crops play a major role in alleviating the malnutrition problem as they are rich source of minerals, vitamins, fibre and contain a fair amount of protein as well as carbohydrates. In addition to local market demand, vegetables have the potential for both domestic and export market. The area, production and productivity under vegetable crops of our country during 1950-51 was merely 2.84 million hectares, 16.5 million tonnes and 5.8 t/ha, respectively and now there has a phenomenal increase in area (3.63 folds), production (10.62 folds) and productivity (2.93 folds) of vegetables during the last seven decades. Presently, vegetables contribute 41.32 per cent area and 59.32 per cent production in total horticultural crops in India. As per second advance estimates of National Horticulture Board for the year 2016-17, the total area under vegetables is 10.31 million hectares with a production of 175.19 million tonnes and productivity of 17 t/ha. Although India is the second largest producer of vegetables next only to China in World, the productivity of different vegetables in our country is comparatively lower than the World's average productivity. More than 40 kinds of vegetables belonging to different groups, namely, solanaceous, cucurbitaceous, leguminous, cruciferous (cole crops) and tuber crops are grown in India.

Yield Losses and Recent Changes in Pest Scenario

A number of insect-pests cause economic losses to the vegetable crops. On an average, the insect-pests inflict crop losses to the tune of 30-40 per cent in vegetable crops (Srinivasan, 1993) and in many cases, there is 100 per cent yield loss due to viral diseases vectored by insect-pests (Shivalingswami *et al.*, 2002). In changing climate scenario, changes in the cropping systems and introduction of highly input responsive high yielding varieties/hybrids; a shift in insect-pest status has been observed. Most of the plant protection recommendations in vegetables indicated the calendar based application of insecticides. This has become a common practice over the years by most of the farmers throughout the country. This malpractice has resulted in developing resistance to several pesticides and often there are secondary out breaks. In addition, many insect-pests have adapted new hosts.

Some of the insect-pests of vegetable crops are gradually attaining the major pest status in different regions of the country due to changes in the ecosystem and habitats. The serpentine leaf miner (*Liriomyza trifolii*) attacks tomato, brinjal, melons, leafy vegetables and cucurbits while the spiralling white fly damages okra. The mealy bug, *Coccidohystrix insolita* and leafhopper, *Empoasca motti*, attack brinjal and bitter gourd, respectively, (Balasubramanian, 2004) cabbage leaf webber, *Hellula undalis* and red spider mites attck on okra, brinjal and cucurbits. In recent times, there were outbreaks of gall midge (*Asphondylia capparis*), which was known to be a minor pest and gradually becoming a regular problem in chilli and capsicum and brinjal in Andhra Pradesh and Karnataka and in brinjal in Chattisgarh. Different species of mealy bugs in cotton, vegetables and papaya have intensified their severity of occurrence in different parts of the country and have become indicator insects for the current ecosystems due to slow changes in climate in the last one decade. Several national and international polyphagus pests like termite, white grub, hairy caterpillar, gram pod borer and tobacco caterpillar are also becoming severe and expanding their host-horizon (Rai *et al.*, 2014).

Injudicious Use and Misuse of Pesticides

Vegetable growers by and large solely depend on chemical pesticides to combat the insect-pests' menace. It accounts for 13-14 per cent of total pesticides consumption as against 2.6 per cent of cropped area (Sardana, 2001). The continuous, excessive and injudicious use of pesticides has triggered several ecological consequences like resistance (currently, more than 500 species of insects) and resurgence of insect pests, pesticide poisoning, environmental pollution, destruction and elimination of natural enemies, effect on non-target organisms, disruption in food cycle and food web and residues in food commodities (Balasubramanian, 2004). It is of huge importance in developing countries where billions of people are living and working in farming areas are exposed to pesticides. Farmers spray methyl parathion over cauliflower to give extra white appearance. Okra is dipped in copper sulphate solution to obtain lush green colour of the fruits. In addition, to obtain shiny skin of tomato and brinjal, the fruits are dipped in the solution of carbaryl (Balasubramanian, 2004).

Pest Management Approaches

Looking in to the present scenario of changing insect-pests' status and the health hazards due to over and misuse of pesticides, there is a need to relook into the production and protection strategies in vegetable based cropping systems. In this context, Integrated Pest Management (IPM) system needs to be intensified for sustainable crop protection strategy against the backdrop of increasing pesticide use and deleterious effect of residues in the environment. IPM system manages the insect-pest population in such a manner that economic loss is avoided and adverse side effects of chemical pesticides are minimized. Monitoring of insect-pests is one of the important components of IPM to take proper decision for the management of insect-pest problem well in time. It can be done through agro-ecosystem analysis (AESA), field scouting, light, sticky and pheromone traps. The economic threshold level (ETL) of important pests and disease are determined to take appropriate control measures when pest population crosses ETL. Certain pests are recorded as the pests of national significance (DPPQS, 2003). AESA is an approach in which extension functionaries and farmers to analyze the field situations in reference to pests, defenders, soil conditions, plant health, the influence of climatic factors and their relationships for growing healthy crops. Such a critical analysis of the field situation will help in taking appropriate decision on management practices.

Table 16.1: Economic Threshold Level of important Insect-Pests of Vegetable Crops

Insect-pests	*Economic Threshold Level*
DBM of cabbage (*P. xylostella*)	2 larvae/plant at 1-4 weeks after transplanting or 5 larvae/plant at 5-10 weeks after transplanting
Cauliflower aphids (*B. brassicae*)	30 aphids/plant
Chilli mites (*P. latus*)	Single mite/leaf
Chilli thrips (*Scirtothrips dorsalis*)	2 thrips/leaf
Whitefly (*B. tabaci*) in brinjal	5-10 flies/leaf
BSFB (*L. orbonalis*)	0.91 per cent on shoots and 0.67 per cent on fruits
Tomato Fruit borer (*H. armigera*)	8 eggs/15 plants or 1 larva/plant or 2 per cent fruit damage
Okra fruit borer (*E. vittella*)	5.3 per cent of fruit infestation
Okra leafhopper (*A. biguttula biguttula*)	4-5 nymphs/plant
Cucurbit leafhopper (*A. biguttula biguttula*)	2.5 nymphs/leaf
Cucurbit epilachna beetle (*Epilachna* spp.)	1 adult/m^2
Whitefly (*B. tabaci*) in okra	4 adults/leaf
Pea aphids (*Acyrtosiphon pisum*)	3-4 aphids/stem tip
Onion thrips (*Thrips tabaci*)	13-15 thrips/plant 15 days after transplanting

Sources: Shivalingaswamy *et al.*, 2002; Singh and Singh, 2002.

DPPQS, 2003;

The importance of host plant inherent resistance or partial resistance as a key component in IPM is increasingly recognized (van Emden, 1991). Host plant resistance to insects is an underutilized pest management strategy in vegetable production. In comparison to cereal and forage based cropping systems, there is much less use of this method for the management of insect-pests in commercial vegetable production (Smith, 1989).

Table 16.2: Tolerant Varieties of Important Vegetables in India

Crops	*Insect-pests*	*Varieties*
Tomato	Fruit borer (*H. armigera*)	ArkaVikash, Pusa Gaurav, Pusa Early Dwarf, Punjab Keshri, Punjab Chhuhara, Pant Bahar, Azad
Brinjal	Shoot and fruit borer (*L. orbonalis*), Aphid, jassid, thrips, whitefly	SM 17-4, PBR 129-5 Punjab Barsati, ARV 2-C, Pusa Purple Round, Punjab Neelam, Kalyanpur-2, Punjab Chamkila, Gote-2, PBR-91, GB-1, GB-6
Cabbage	Aphid (*Brevicoryne brassicae*)	All season, Red Drum Head, Sure Head, Express Mail
Cauliflower	Stem borer (*Hellula undalis*)	Early Patna, EMS-3, KW-5, KW-8, Kathmandu Local
Okra	Jassid (*Amrasca biguttula biguttula*)	IC-7194, IC-13999 New Selection, Punjab Padmini
Onion	Thrips (*Thrips tabaci*)	PBR-2, PBR-6, Arka Niketan, Pusa Ratnar, PBR-4, PBR-5, PBR-6
Round gourd, Pumpkin, Bitter gourd	Fruit fly (*B. cucurbitae*)	Arka Tinda, Arka Suryamukhi, Hisar-II
	Red pumpkin beetle (*Aulacophora* spp.)	NB-19, NB-25, NB-29

Sources: Dhandapani *et al.*, 2003.

Nath and Thakur, 1965.

There are several agronomic practices that make the environment less favourable to the insect-pests. Crop rotation, location and alteration in sowing/ planting date reduce the attack of red pumpkin beetle, fruit fly, shoot and fruit borer. Early planting of cucurbits in November escapes the attack of red pumpkin beetle whereas, flowering beyond October in bitter gourd suffers less from fruit fly. Sowing of okra during second week of June retain less population of borers, thereby incurred maximum healthy yield whereas, July planted brinjal face more attack of shoot and fruit borer. Thus, synchronization of most susceptible stage of the crop with the inactive period of insect reduces the infestation and chemical intervention. Besides, summer ploughing is an effective practice to destroy the soil dwelling inactive stages of fruit fly, red pumpkin beetle and cut worm as they are exposed for sun drying and for birds and other predators.

Trap crops are planted to attract and hold insect-pests where they can be managed more efficiently and reduce their movement on to the major crop. Trap crop must be distinctly attractive to the pest than the main crop. It provides protection either by preventing the pest from reaching the main crop or by concentrating them in certain part of the field where they can be economically destroyed. Mustard and Chinese cabbage as a trap crop and coriander and tomato as deterrent crop along with cabbage has been successfully utilized for the control of diamondback moth, aphid and leaf webber (Srinivasan and Moorthy, 1992 and Satpathy *et al.*, 2010). Giant African marigold functions as good source to trap the adults of *Helicoverpa armigera* besides it also attracts the adults of leafminer for egg laying on the leaves. Maize plants applied with bait spray trap and kill fruit fly adults when sown in combination with bitter gourd (Shooker *et al.*, 2006). Yellow vein mosaic disease can be minimized by trap cropping of corn in okra fields (Adhikary *et al.*, 2015).

The conservation of natural enemies is probably the most important and readily available biological control practice available. Natural enemies occur in all vegetable production systems. They are adapted to the local environment and to the target pest and their conservation is generally simple and cost effective. Lacewings, lady beetles, syrphid fly larvae, and parasitized aphid mummies are almost always present in aphid colonies. Fungus infected adult flies are often common during periods of high humidity. These natural controls are important and need to be conserved when making pest management decisions. Inundative release of *Trichogramma brasiliensis* @ 240000/ha are also recommended for control of fruit borers on okra and tomato (Jalali *et al.*, 2005). Six releases of *T. brasiliensis* at weekly interval @ 40000/ha the first release coinciding with 50 per cent flowering and thereafter at weekly interval in tomato is recommended. *Chrysoperla zastrowiarabica* is an effective predator for control of whitefly, aphid, jassid and eggs of some lepidopterous borers when the first instar larvae are released @ 50000/ha (Khulbe *et al.*, 2005). *Cotesia plutellae* and *Diadegma semiclausum* can be incorporated in to biological pest management of diamondback moth as larval parasitoids (Saucke *et al.*, 2000). Microbial biopesticides are eco-friendly and compatible with natural enemies existing in the ecosystem. It has been widely applied against *Plutella xylostella, Earias vitella* and *H. armigera.* Because of host specificity, use of Nuclear Polyhedrosis Virus (NPV) is limited to *H. armigera* and *Spodoptera litura.* Lack of commercial production has suppressed its popularity and is used in limited areas.

Insecticides are needed when other management tactics fail to keep pest population below the level that causes economic damage. The choice of insecticides, formulations, and rate and method of application will vary depending on several factors, including the insect-pest(s) to be controlled and the stage of crop development. The different steps are recommended to minimize the development of resistance to insecticides: (a) apply insecticides only when needed (b) follow recommended action thresholds where available (c) use the minimum recommended rate that provides control (d) rotate the insecticides that are in different chemical classes and have different modes of action during the season and (e) use effective insecticide synergists if available. The insect-pest management through chromatic traps, mechanical traps, light traps, sticky traps and pheromone traps may give an added advantage. In addition, semio-chemicals (behavior-modifying compounds) and bio-pesticides (microbial control agents including entomo-pathogenic fungi, bacteria and viruses and botanical bio-pesticides) are gaining importance particularly in insect-pest management of vegetable crops.

References

Adhikary, S.; Pandit, M.K.; Koundinya, A.V.V.; Bairagi, S. and Das, A. 2015. Examination of system productivity and profitability of baby corn based vegetable intercropping systems. *Journal Crop and Weed* 11(1): 220-224.

Balasubramanian, P. 2004. The Vegetable Industry in India: Critical IPM Needs. The Asian Regional IPM CRSP Workshop- Impact and lessons learned. June 29-30, 2004, Philippines.

Dhandapani, N.U.; Shekhar, R. and Murugan, M., 2003, Bio-intensive pest management (BIPM) in major vegetable crops: An Indian perspective. Food, Agric. Environ., 1:333-339.

DPPQS, Directorate of Plant Protection, Quarantine and Storage, 2003. Integrated Pest Management Package, Government of India. (IPM Package No: 18, 21, 2223).

Jalali, S.K.; Murthy, K.S.; Venkatesan, T.; Lalitha, Y. and Devi, P.S. 2005. Preliminary testing of some new release methods for egg parasitoid *Trichogramma* spp. *Journal of Biological Control* 19(2): 99-103.

Khulbe, P.; Maurya, R.P. and Khan, M.A. 2005. Biology of *Chrysoperla carnea* (Stephens) on different host insects. Annals of Plant Protection Sciences 13(2): 351–354.

Nath, P. And Thakur, M.R. 1965. Evaluation for red pumpkin beetle resistance in gourds. *Indian Journal of Horticulture*, 22:330-331.

Rai, A.B.; Halder, J. and Kodandaram, M.H. 2014. Emerging insect pest problems in vegetable crops and their management in India: An appraisal. Pest Management in Horticultural Ecosystems 20 (2): 113-122.

Sardana, H.R. 2001. Integrated pest management in vegetables. In: Training manual -2, Training on IPM for Zonal Agricultural Research Stations. May 21-26, 105-118 pp.

Satpathy, S.; Shivalingaswamy, T.M.; Kumar, A.; Rai, A.B. and Rai, M. 2010. Potentiality of Chinese cabbage (*Brassica rapa* subsp. *pekinensis*) as a trap crop for diamondback moth (*Plutellaxylostella*) in cabbage. *Indian Journal of Agricultural Sciences* 80(3): 238-241.

Saucke, H., Dori F. and Schmutterer, H. 2000. Biological and Integrated Control of *Plutella xylostella* (Lep., Yponomeutidae) and *Crocidolomia pavonana* (Lep., Pyralidae) in Brassica Crops in Papua New Guinea. *Biocontrol Science and Technology* 10(5):595-606.

Shivalingaswamy, T.M.S.; Satpathy, S. and Banerjee, M.K. 2002. Estimation of crop losses due to insect pests in vegetables, In: Resource Management in plant protection, Vol-I (SaratBabu, B.; Vara Prasad, K.S.; Anita, K.; Rao R.D.; Chakraborty, S.K. and Chandukar, P.S. eds.) Plant Projection association of India, Hyderabad, pp. 24-31.

Shooker, P.; Khayrattee, F. and Permalloo, S. 2006. Use of maize as a trap crops for the control of melon fly, *B. cucurbitae* (Diptera:Tephritidae) with GF-120. Bio-control and other control methods (online). Available on: http//www.fcla.edu/flaEnt/fe87 p354. Pdf.

Singh, Y.P. and Singh, P.P. 2002. Nature and extent of damage and avoidable yield losses due to shoot and fruit borer (*Leucinodes orbonalis*) to brinjal (*Solanum melongena*) at medium high altitude hills of Meghalaya. *Indian J. Ent.*, 64 (1): 12-16.

Smith, C.M. 1989. Plant resistance to insects. Wiley, New York, 286 pp.

Srinivasan, K. 1993. Pests of vegetable crops and their control. In: Advances in Horticulture, 6: Vegetable crops Eds. K.L. Chadha and G. Kalloo. Malhotra Pub. House, New Delhi. p. 859-886.

Srinivasan, K. and Moorthy, P.N.K. 1992. Development and adoption of IPM for major pests of cabbage using Indian mustard as trap crop. In: Diamondback moth and other crucifer pests: Proceedings of 2nd International Workshop, Taiwan, 10-14 December, 1990.

Van Emden, H.F. 1991. The integration of host plant resistance to insects with other control measures, pp. 939-948. *In* Proceedings of the Brighton Crop Protection Conference on Pests and Diseases -1990, Vol. 3. British Crop Protection Council, Surrey United Kingdom, 474 pp.

Chapter 17

Increasing Farmers' Income through Technological Interventions in Vegetable Production

A.C. Mishra, S.V. Diwedi, R.K Singh, S. Kumar, Neetu and Sweta Soni

Department of Vegetable Science, Banda University of Agriculture and Technology, Banda, U.P.

Introduction

Vegetables are important constituents of Indian agriculture and nutritional security due to their short duration, high yield, nutritional richness, economic viability and ability to generate on-farm and off-farm employment. Our country is blessed with diverse agro-climates with distinct seasons, making it possible to grow wide array of vegetables. India is the second largest producer of fruits and vegetables in the world. Total area under horticultural crops is 21.83 million hectares and production is 240.53 million tones. Fruits and vegetables together contribute about 92 per cent to the total horticultural production in the country. India produces 14 per cent (146.5 million tonnes) of world's vegetables on 15 per cent (8.5 million hectares) of world area under vegetables. Productivity of vegetables in India (17.3t/ha) is less than the world average productivity (18.8t/ha). Potato (28.9 per cent), tomato (11.3 per cent), onion (10.3 per cent) and Brinjal (8.1 per cent) are the 4major vegetables contributing 58.6 per cent of total vegetable production in our country. Other important vegetables are cabbage (5.4 per cent), cauliflower (4.6 per cent), okra (3.9 per cent) and peas (2.4 per cent). India ranks first in the production of okra in the world (73 per cent of world production).

The country has witnessed tremendous progress in vegetable production, especially during the post green revolution period. Development of improved vegetable varieties/hybrids/technologies through systematic research coupled with their adoption by the farmers and developmental policies of the government culminated in tremendous increase in area under vegetables (8.5 million ha), production (146.5 million t) and productivity (17.3 t/ha) in the country. Compared to area (2.84 million ha), production (16.5 million t) and productivity (5.8 t/ha) in 1950-1951, there had been phenomenal increase in area (2.99 folds), production (8.88 folds) and productivity (2.98 folds) of vegetables in our country during the last 6 decades. The per capita availability of vegetables is also on the increasing trend from 147.82 g/capita/day in 1991 to 230.4g/capita/day in 2011. Increasing affordability, health consciousness, urbanization, involvement of women force, market demand for high value vegetables, favourable income elasticity of demand and growth rate for domestic consumption of fruits and vegetables are some of important reasons for vegetable growth in the country (Vanitha *et al.*, 2013).

Now a day, increased production and productivity is scanned under the frame of profitability of the enterprise. Being a short duration crop, vegetables are the most suitable for increasing crop intensity and income in subsistence farming. In addition to increasing productivity, off season cultivation, resource conservation, post harvest processing and value addition and integration of animal component are also utmost important for increasing profitability in vegetable cultivation. Some of the researched based findings are discussed her to illustrate the spirit of the subject.

1. Increasing the Income through Vegetable Based Inter cropping Systems

Inter cropping is the way to harvest the residual resources without any adverse effect on main crop and thus, it is ever beneficial than sole cropping. Many of the compatible companion crops have been found to give synergic effect on overall profitability of the production system.

a. Maize + Pole type Cowpea System

Some of the maize + vegetable intercropping systems have also been evaluated for working out the profitability and it was found that maize + pole type cowpea in 1:1 ratio was the best system in sense of yield of main crop (maize) and intercrop (cowpea), maize equivalent yield and B:C ratio followed by maize + Ridge gourd var. *Satputia* (1:1) (Table 17.1). The lucrative outcome from such intercropping systems promoted the farmers to maintained proper spacing in maize which lead to higher yield of this crop. This also forced the growers to realize the importance of proper spacing in crops in general and particularly in maize. Such participatory experiments had wide effects among the growers to adopt the technology and generate reliance upon the researchers (Mishra, 2016).

Table 17.1: Efficiency of Maize-Vegetables Intercropping Systems in Jharkhand Average of Two Years

Technology Assessed	*Yield/plot (q)*		*Maize-Equivalent Yield (q/ha)*	*B:C Ratio*
	Maize	*Intercrop*		
Maize as sole crop at spacing of 40 cm x 20 cm (as practiced by the farmers)	17.71	-	15.82	1.95
Maize (60 cm x 25 cm) + Pole Cowpea (1:1)	24.64	70.24	44.83	3.80
Maize (60 cm x 25cm) + Ridge gourd var. *Satputia* (1:1)	19.25	58.26	41.60	3.60
Maize (60 cm x 25 cm) + Cucumber (1:1)	19.42	13.85	29.05	2.50
CV (per cent)			9.74	
SE (m)			1.01	
CD (at 0.05)			2.07	

Note: Average sale rate of Cowpea Rs. 200/q^{-1}, Ridge gourd Rs. 300/q^{-1}, Cucumber Rs. 500/q^{-1} and maize Rs. 800/q^{-1}

b. Pigeonpea + Okra Systems

Pigeonpea (Arhar) is extensively grown pulse crop in Jharkhand particularly in poorly fertile, degraded rainfed uplands as sole or mixed crop with Sesame. Although, Pigeonpea suffers a large number of diseases and insect-pests and engages the land for almost whole year, the popularity of this crop in all the way is stagnant in sense of area and trade. This is because of extensive area under rainfed which would remain uncultivated if not covered under Pigeonpea. The Pigeonpea suffers a number of biotic threats like wilt, sterility mosaic disease and pod borer insects in Jharkhand. Apart from biotic challenges, the crop production in this area is still done through non-scientific traditional systems using untreated traditional seeds, without application of manures and fertilizers and plant protection chemicals. Furthermore, most degraded, dried, uncultivated and nutrient-starved lands are used for Pigeonpea cultivation in this state. All these lead poor productivity of the crop. The Agricultural Institutes in the state have recommended many technological interventions to improve the productivity of the crop *viz.*, high yielding varieties (Narendra Arhar-1, Birsa Arhar-1, Malviya-13, Bahar *etc.*), Seed treatment with fungicides or *Trichoderma*, application of *Rhizobium* biofertilizer and proper NPK, line sowing at proper spacing and plant protectants for controlling diseases and insects but availability of resources at inconvenient place and time, poor conduction of information among the growers and poor socio-economic condition of the peasantry lead poor adoption of the technologies.

As a matter of perception and mindset, it was worked out that the growers would practice fertilization and intercultural operations in Pigeonpea crop only when some other crop is associated with this crop. Therefore, some most common intercrops like sorghum, sesame, groundnut and okra were planned to grow with pigeonpea for assessment of net return likely to come (Table 17.2). It was found that pigeonpea var. *Bahar* sown at 75 x 30 cm spacing + okra cv. Arka Anamika gave maximum Pigeonpea equivalent yield (22.46 q/ha) and B:C ratio (3.4).

Table 17.2: Efficiency of Pigeonpea-Based Intercropping Systems in Jharkhand

Technology Assessed	*Yield (q/ha)*		*Pigeon Pea Equivalent Yield (q/ha)*	*B:C Ratio*
	Main Crop	*Inter Crop*		
Mixed sowing of Pigeonpea var. Bahar and Sesame var. Kanke Safed (as practiced by the farmers)	12.2	1.45	13.40	1.6
Pigeonpea var. Bahar (75 x 30 cm) + Sorghum var. CSV-20 (1:1 Ratio)	14.3	17.2	18.89	2.8
Pigeonpea var. Bahar (75 x 30 cm) + Okra cv. Arka Anamika (1:1 Ratio)	14.5	47.73	22.46	3.4
Pigeonpea var. Bahar (90 x 30 cm) + Groundnut var. AK-12-24 (1:3 Ratio)	13.2	13.7	20.05	3.0
SE (m)			1.6	
CD (at 0.05)			3.28	

Note: Selling rate of Pigeonpea Rs. 3000/q^{-1}, Sesame Rs. 2500/q^{-1}, Okra Rs. 500/q^{-1}, Sorghum Rs. 800/q^{-1}, Groundnut Rs. 1500/q^{-1}

c. Okra + Pole Type Cowpea System

Vegetable cultivation is an important enterprise of small and marginal famers in irrigated lands. In general, two three sole crops of different vegetables are taken from same field in a year depending upon the availability of irrigation water. Many times sole crop does not ensure profitable yield particularly in the case of the crops vulnerable to epidemic diseases like late blight in potato and tomato, yellow vein mosaic and leaf curl diseases. In such circumstances intercropping provides a safeguard from complete loss of crop. Farmers frequently intercrop radish, broad bean, vegetable pea, rye/mustard or wheat with potato, radish, palak or coriander with tomato and brinjal and amaranth with cucurbits. Two crops of okra are taken, one in summer and another in rainy season. The summer crop remains free from major diseases and insects but that grown during rainy season has major risk of yellow vein mosaic disease and fruit borer insects. Intercropping of rainy okra with compatible crops ensures marginal gain from the crop which would otherwise lead to complete loss with the biotic stresses. In an experiment, two intercrops cowpea (pole type) and ridge gourd (*Satputia*) was compared with sole cropping of okra and it was found that okra + pole type cowpea in 1:1 ratio was most remunerative in terms of okra equivalent yield (135.17 q/ha) and B:C ratio (4.28). Ridgegourd (*Satputia*) did not exhibit compatibility with okra because of its vigorous vining nature leading to poor fruiting in okra (Table 17.3) (Mishra, 2013).

2. Resource Conservation

Input cost in cultivation and market chain and demand decide ultimate income from an agricultural enterprise. Optimum use and efficient utilization of resources are the crucial aspect for increasing profit. Expenses on seed, fertilizer, irrigation water and inter-cultural operations should be well governed. The concept of precision farming is highly valid in this era of commercial farming. The effects of plastics for nursery raising, irrigation and controlling edaphic and climatic environment have been well understood for precision farming of vegetable crops.

Table 17.3: Efficiency of Okra-based Intercropping Systems in Jharkhand

Technology Assessed	*Yield. (q/ha)*		*Okra Equivalent Yield (q/ha)*	*B:C Ratio*
	Okra	*Intercrop*		
Sole crop of okra at 30x15cm spacing (as practiced by the farmers)	98.45	-	98.45	3.63
Okra at recommended spacing (45x30 cm) intercropped with pole type Cowpea in alternate rows (2:1 ratio)	87.58	47.54	135.17	4.28
Okra at recommended spacing (45x30 cm) intercropped with Ridge gourd cv. *Satputia* in alternate rows (2:1 ratio)	53.60	56.57	96.023	3.23
CV (per cent)	-	-	5.69	-
SE (m)	-	-	1.98	-
CD (at 0.05)	-	-	4.16	-

Note: Selling rate of okra Rs.400/-q^{-1}, Ridge gourd Rs.300/q^{-1} and cowpea Rs.400/q^{-1}

a. Paired Row Planting Geometry in Potato

The traditional and scientifically recommended system of potato cultivation involves planting of seed tubers of 40-50g weight at 60x20 cm spacing in single row per ridge. The paired row planting geometry consisted of planting the seed tubers weighing 20±5g in 15 cm apart paired-rows on ridges spaced at 60x20 cm (centre to centre) accompanied with furrow irrigation and it was compared with traditional planting systems (single row) at variable spacing (Mishra, 2013). In another experiment, the paired-row planting geometry was compared with traditional practices comprising planting of seed tubers of white tubered cultivars *Kufri Himalini* and *Kufri Girdhari* on the ridges with single row at 60x20 cm planting distance along with drip and/or furrow irrigation systems (Mishra and Pandey, 2016).

Planting the seed tubers of 20±5 g in paired rows 15 cm apart on the ridges made at 60 cm isolation (centre to centre), resulted in 244.9 q/ha tuber yield in potato cv. *Kufri Kanchan* as compared to recommended planting system of single row at 60x20 cm spacing using seed tubers of 50-60g (173.5 q/ha) and farmers' practice *i.e.* single row at 45x20 cm (129.6 q/ha). The seed requirement in traditional system and paired system remains the same as smaller seed tubers are used in the case of latter (Mishra, 2013).

Two varieties of potato Kufri Girdhari and Kufri Himalini responded well to paired row panting geometry. Among the cultural packages paired row planting + furrow irrigation was most promising combination for tuber yield (35.2 t/ha) followed by paired row planting + drip irrigation (33.1 t/ha) in temperate Himalayas (Mishra and Pandey, 2016).

b. Use of Plastic Mulches and Drip Irrigation

Mulching with black polythene, white polythene and even dries grasses have been found to give drastic increase in yield due to conservation of soil moisture, regulation of soil temperature and suppression of weeds in hilly areas as well as plains. Polythene mulching accompanied with drip irrigation system have been found even more efficient in terms of water resource conservation. In temperate hills of Uttarakhand, mulching with black and white polythene (100 μ) and 10 cm thick bed of dried grasses accompanied with or without drip irrigation were tested in marrow (*Cucurbita pepo*) crop. Results indicated that drip irrigation was consistently superior in relation to fruit yield (694.1 q/ha) over traditional system of basin irrigation (594.5 q/ha) over two years. Mulch application resulted in 44.9-47.2 per cent increase in fruit yield over the unmulched crop. Higher and *at par* fruit yield was realized by application of dried grass (705.1 q/ha) and black polythene mulches (704.9 q/ha). The interaction two treatment combinations indicated that black and white polythene mulches accompanied with drip irrigation exhibited high and *at par* fruit yield in marrow (*C. pepo*) over the years (797.6 q/ha and 788.5 q/ha, respectively) (Mishra, 2017).

Similarly, black and white polythene mulches were tested by Mishra (2018) in brinjal during summer-rainy seasons against unmulched treatment. Both type of polythene had enthusiastic results in regulation of soil temperature and fruit yield. White polythene owing to green house effect raised soil temperature by almost 8 °C and four times higher fruit yield (446.30 q/ha) as compared to unmulched control (24.9°C and 112.22 q/ha, respectively) (Table 17.4).

Table 17.4: Effect of Polythene Mulching on Fruit Yield of Brinjal during Summer-Rainy Season

Sl. No.	*Mulching*	*Mean Soil Temperature (°C)*	*Biomass of Weeds/m²(g)*	*Number of Fruits per Plant*	*Fruit Yield per Plant (kg)*	*Fruit Yield per Hectare (q)*
1.	Black Polythene	30.6	45.8	20.41	1.19	330.37
2.	White Polythene	33.4	750.6	25.79	1.61	446.30
3.	Without mulching	24.9	2675.5	8.51	0.404	112.22
	CV (per cent)	9.3	16.4	7.5	11.2	13.2
	CD (0.05)	0.9	175.3	2.6	0.4	39.6

Drip irrigation has also been tested in onion under different doses of NPK and it was found that drip irrigation had significantly higher number of leaves per plant (9.95), leaf dry matter content (43.56 per cent), lateral and vertical bulb diameter (6.31 cm and 8.62 cm, respectively) and bulb weight (131.49 g). However, bulb yield, specific gravity of bulbs and bulb volume was not affected significantly by irrigation system although higher values were recorded in drip system for bulb yield (403.3 q/ha) and specific gravity of bulbs (0.91 g/cm³). Three levels of NPK applied had significant influence on bulb yield. There was parallelism in increasing doses of NPK from NPK @50:40:40 kg/ha to NPK @ 150:80:60 kg/ha and bulb yield from 353.0 to 441.10 q/ha. The combined effect of irrigation systems and NPK levels had also significant impact on the all bulb yield and plant growth characters. Maximum bulb yield was registered in drip irrigation accompanied with NPK level of 150:80:60 kg/ha (453.30 q/ha) followed by sprinkler irrigation accompanied with NPK level of 150:80:60 kg/ha (428.90 q/ha) which were statistically *at par*. In the areas having limited irrigation water, drip system is equally effective for increasing bulb yield through higher level of NPK. On the basis of above findings, irrigation through drip and sprinkler and application of NPK @150:80:60 kg/ha could be recommended for higher bulb yield in onion (Mishra and Raturi, 2015).

c. FYM doses and Black Polythene Mulching

The effect of two FYM levels *i.e.* 7.5t/ha and 15 t/ha along with or without black polythene mulch (200 micron) was evaluated in sponge gourd line BUAT SG18-1 (IC 0628847) during rainy season, 2018 in RBD with five replications. Application of FYM @15t/ha with black plastic mulching was found to be best treatment with 223.9 q/ha fruit yield and B/C ratio of 5.9 in sponge gourd BUAT SG 18-1 (IC 0628847) during rainy season. The results on fruit yield and economics has been given as under:

Treatment Details	*Number of Fruits/ Plant*	*Average Fruit Weight (g)*	*Average Fruit Length (cm)*	*Fruit Yield (q/ha)*	*Cost of Cultivation (Rs./ha)*	*B/C Ratio*
FYM @7.5t/ha without plastic mulching	10.5	84.6	21.6	44.5	38,650	2.3
FYM @15t/ha without plastic mulching	18.8	96.4	23.2	90.6	46,150	3.9
FYM @7.5t/ha with black plastic mulching	26.6	117.8	29.6	156.7	68,650	4.6
FYM @15t/ha with black plastic mulching	35.4	126.5	33.4	223.9	76,150	5.9
CV (per cent)	11.3	13.2	12.7	14.8	-	-
CD (0.05)	3.6	9.8	4.7	67.5	-	-

Annon. 2019.

3. Integration of Animal Components in Cereal and Vegetable-Based Farming

In general, animal husbandry (cattle, poultry, pigs, goats and sheep) is much important for vegetable cultivation in the view of higher nutrient required to the crops is supplemented through the manures. Some of the animal based enterprises like piggery is even more compatible with vegetable cultivation in the view of residue recycling. Vegetable residues could be used to supplement pig feed in a big way. On the basis of five years observations it was found that piggery of T and D breed was found the most compatible, viable and profitable component in baseline farming system including cereal crops + vegetables +poultry and duckery + fishery and cereal crops + vegetables + poultry in terms of net profit from piggery (Rs. 1,60,000/unit/year and Rs. 1,54,000/unit/year, respectively), net profit from farming system (Rs. 2,23000/year and Rs. 3,05,000/year, respectively) and B/C ratio (5.7 and 5.5, respectively) (**Mishra and Baxla, 2015**)

4. Off Season Cultivation of Vegetables

The objective off season cultivation is to produce vegetables beyond its normal season so that higher value of the produce could be procured. Cultivation of vegetable crops in green house structures, river bed cultivation, rainy season cultivation of tomato and cucurbits are the some of the examples of off season production. Specific package of practices is required for such type of enterprises.

5. Introduction of High Yielding Varieties

High yielding and stress tolerant varieties have been found to be helpful in restoring sustaining the yield and income of farming community. Commercialization of new short duration crops like mustard greens (*Brassica juncea* L.) and popularization of stress tolerance varieties in Bundelkhand region have been found as a mean for raising farmer's income. Some of the examples of high yielding and drought tolerant genotypes of are as follows:

Sponge gourd- BUAT SG18-1 (IC 0628847)

Plants tolerant to moisture stress and spreading type. Leaves 5 lobed, flattened, male and female flowers start from 5-6 and 8-9 nodes, respectively. Fruits light green, slender, 25-30 cm long, 120-125g in weight, 30-35 fruits per plant and fruit yield 230-250 q/ha under proper crop management. Fruits give pleasing aroma and taste on cooking. This suitable for cultivation in both summer and rainy season.

Mustard Greens- BUAT MG-1

This is selection from IC 0612094. Leaves broad, flattened, dark green, crisp with mild pungency. Leaf picking starts from 30-55 days after sowing. The green leaf yield of 640-650 q/ha from direct sown crop at 30 x 10 cm spacing.

6. Post-harvest Processing and Value Addition

The technological interventions to induce productivity or reduce input cost may lead to higher profit from farming. However, secondary agriculture comprising processing and value addition have even more potential to raise income from farming. Therefore, infrastructural setups are very important for increasing the farm income.

References

Anonymous, 2019. BUAT Annual Reports-2019.

Mishra A.C. and Raturi A. 2015. Effect of irrigation systems and NPK levels on onion (*Allium cepa* L.) in temperate hills. *Vegetable Science*, 42(2): 107-108.

Mishra, A.C. 2013. Increasing Profitability and Resource Conservation through Paired-Row Planting Geometry in Potato. *Potato Journal*, 40(2): 180-183.

Mishra, A.C. 2016. On farm verification of maize-vegetable based intercropping systems under subsistence farming of Jharkhand. *Vegetable Science*, 43(1):135-137.

Mishra, A.C. 2017. Effect of irrigation systems and mulching on soil temperature and fruit yield of marrow (*Cucurbita pepo*) in temperate Himalaya of Uttarakhand. *Vegetable Science*, 44(1): 66-70.

Mishra, A.C. 2018. Effect of plastic mulches on performance of brinjal (*Solanum melongena* L.) in temperate Himalaya. *Vegetable Science*, 45(2): 300-302.

Mishra, A.C. and Baxla, S.L. 2015. Compatibility and productivity of Piggery in different farming modules in Jharkhand. *The Bioscan*, 10(2):781-784.

Mishra, A.C. and Pandey, V. 2016. Performance of potato varieties under different moisture management systems and planting geometry. *The Asian Journal of Horticulture*.11(1): 154-158.

Mishra, AC. 2013. Efficiency of okra-based intercropping under subsistence farming. *Vegetable Science* 40(1):105-106.

Vanitha, S.M.; Chaurasia, S.N.S.; Singh, P.M. and Naik, P.S. 2013. Vegetable Statistics. *Technical Bulletin* No. 51, IIVR, Varanasi, pp. 250.

Chapter 18

River Bed Cultivation

S.V. Dwivedi, Neetu, Sunil Kumar, Shweta
A.C. Mishra and R.K. Singh

Department of Vegetable Science,
Banda University of Agriculture and Technology, Banda, U.P.

River bed commonly known as diara land is a basin or bank area or area between two or more streams of river which seems like lighted Deepak or Diya.

River bed cultivation is a kind of vegetable forcing of off season vegetable production technology which is purely an indigenous and innovation of Indian vegetable growers. In India and abroad, history of diara land is not in record but the archeological studies have indicated river bed cultivation is being practiced in India since ancient times on the basin of rivers after water level receeded.

In India, cultivation of cucurbits like watermelon, muskmelon, longmelon, pumpkin, summer squash, ridge gourd, smooth gourd, snake gourd and snapmelon mostly in North and Central India, bitter gourd, snake gourd *etc.* in some areas like Kerala and pointed gourd in Eastern U.P., Bihar and West Bengal is more common under river bed condition.

Presently not only in India, but in other south east Asian countries also, cucurbits are commercially cultivated in the river bed areas. In survey it is observed that out of total area under cucurbits cultivation 60 per cent area is under river bed cultivation and during summer season around 75-80 per cent of total cucurbits production are being produced in diara land area which is available in the market from February to June (Ram *et al.*, 2007).

It is an essentially a mixed cropping system with phased sowing to reach the vegetable markets successively with bottle gourd, summer squash, muskmelon, watermelon and lastly pumpkin from February of June, because mixed cropping has some advantages for the growers as it generates continuous source of income from February to June. Generally, socially and economically backward farmers practice this system with in north India, there is some kind of ethnic association with Muslims. This system is mostly practiced in meandering river beds in far flung and in accessible regions (Sehardri and Chatterjee, 2000). A list of commercially cultivated cucurbits under river bed system with their common and botanical names and chromosome number are given in Table 18.1.

Table 18.1: List of Cucurbits Cultivated under River Bed System

Hindi Name	*English/ Common Name*	*Botanical Name*	*Chromosome No.*	*Edible Portion*
Lauki/Ghiya	Bottle gourd	*Lagenaria Siceraria*	22	Endocarp
Karela	Bitter gourd	*Memordica charantia*	22	Endocarp
Kali Tori/Taroi	Ridge gourd	*Luffa acutangula*	26	Endocarp
Chikani Tori/Nenuna	Sponge gourd	*Luffa cylindrical* (*Syn. L. aegyptiaca*)	26	Endocarp
Kharbuja	Muskmelon	*Cucumis melo*	24	Pericarp (with vary little mesocarp)
Tarbooj	Watermelon	*Citrullus lanatus* (*Syn. C. vulgaris*)	22	Placentae
Kakari	Longmelon	*Cumumis melo var. utilissimus* (*Syn. C. melo var. flexosus*)	24	Placentae
Parwal	Pointed gourd	*Trichosanthes dioica*	24	Endocarp
Khira	Cucumber	*Cucumis sativus*	14	Placentae
Sitaphal/Kashiphal	Pumpkin	*Cucurbita moschata*	40	Pericarp (with vary little mesocarp)
Chhappan Kaddu	Summer squash	*Cucurbita pepo*	40	Pericarp (with vary little mesocarp)
Petha	Ash gourd	*Beninijcasa hispida*	24	Endocarp
Phoot	Snapmelon	*Cucumis melo var. momordica*	24	Placentae
Chichinda	Snake gourd	*Tricosanthes cucumerina* (*Syn. L. aegyptiaca*)	24	Endocarp

Classification of Diara Land

Diara land or river bed can be classified into three categories on the basis of their soil composition and cultivation of crops.

1. **Upper Diara:** This group of soil contains clay and silt in major proportion and area is less flooded also. Pointed gourd, bottle gourd, bitter gourd, smooth gourd, ridge gourd and ash gourd are the main crops. Besides, tomato, brinjal and potato are other major vegetables grown by the farmers.
2. **Middle Diara :** This kind of diara contains more sand and less clay. Major crops of such area are ridge gourd, smooth gourd, cucumber, pumpkin,

summer squash, pointed gourd, muskmelon, watermelon and snake gourd.

3. **Lower Diara :** In such areas soil mainly contains sand and is suitable for the crops like watermelon, muskmel9on and longmelon. Cultivation of potato in lower diara of Benas river in Benaskatha district of Gujrat is an exceptional case.

Major, Diara, River Beds of India

In diara/river basin areas, cultivation of cucurbits is a common feature across the country from the river Ganga and Yamuna in North India to Pamba in Kerla.

River Ganga has the maximum area under diara land cultivation, which is estimated to be around 8.65 lakh ha. In Bihar and 2.4 lakh ha in other states and also a most fertile basin area followed by Burhi Gundak with an area of around 2.25 lakh ha., Gandak with 1.45 lakh ha. one with 1.13 lakh ha. and Kosi with an area of 1.45 lakh ha.

Benefits of River Bed Cultivation

The main advantages of this type of cultivation and vegetable production technology are:

- Higher net return per unit area
- Early crop with better quality
- Fresh vegetables are obtained in off season
- Low cost of cultivation
- Better utilization of such cultivable lands
- Lesser use of manures and fertilizers
- Lesser weed infestation
- Easy in irrigation
- Less affected by insects and diseases
- Landless farmers can get river bed land on lease basis from government to get employment and increase their income.
- This is a repository of indigenous germplasm of various cucurbits adopted to off season technology.
- Easy transportation by rivers.
- Availability of adequate water for cleaning of vegetables

Selection of Varieties

There is no variety bred especially for river bed areas, so for. However, number of improved/hybrid varieties have been developed from various public and private sectors, can be utilized to increase production and productivity. The list of important varieties of cucurbits suitable for river bed cultivation is given in Table 18.2.

Table 18.2: List of Important Varieties

Crop	*Open Pollinated Varieties*	*Hybrid Varieties*
Bottle gourd	Narendra Rashmi, Kashi Ganga, Pusa Naveen, Punjab gol, PBOG-61, Arka Bahar, Punjab Komal, Narendra Dharidar, Punjab round, Punjab long	NDBGH-4, Kashi Bahar, Anmol, Lina Varad, Ruma, NS-381, Pant Sankar Lauki-1, Pant Sankar Lauki-2, Miridula, Swati, NOS-204, MGH-1
Bitter gourd	Pusa Vishesh, Pusa Do Mausami, Priya, Arka harit, Punjab-14, Kalyanpur Sona, Pant Kerala-1, Pant Karela-2, Coimbatore long, Kalyanpur Barahmasi	NS-1024, Indira, NBGH-167, Prachi Tijarti, Vivek, Poly, Pusa hybrid-2, Pusa hybrid-1
Ridge gourd	Pusa nasdar, Arka sumit, Arka sujat, Kalyanpur Dharidar, PRG-7, CHRG-1, Co-1, Punjab Sadabahar, Swarn majari, Swarn upahar, HaritHAM	Rohini, Mallika, Naga Gaurav, Surekha, NS-401
Sponge gourd	Pusa Chikni, PSG-40, Pusa Supriya, KSG-14, CHSG-1, Rajendra Nenua-1	Chetak, MSGH-1, Utsav, Harita
Muskmelon	Hara madhu, Durgapur madhu, Pusa Sharbati, Kashi madhu, Arka jeet, Arka rajhans, Pusa madhuras, Hisar madhur, Punjab sunehari	NS-81, MakiNS-81, Makina, NS-17455, MHY-5, M-3, Punjab hybrid-1, Rasika
Watermelon	Sugar baby, Durgapur meetha, Arka manik, Durgapur kesar, Durgapur local	MHW-6, Roja, Madhuri, NS-295, NS-261, NS-246, NS-1004, Tiger, Mukasa, NUN-305, Khusbu, Madhwi, Arka jyoti
Longmelon	Arka sheetal	-
Pointed gourd	Swarn rekha, Swarn alaukik, Kashi alankar, Narendra-260, Narendra-307, Rajendra parwal-1, Rajendra parwal-2	-
Cucumber	Poinsette, Japanese long green, Straight-8, Khira-75, Sheetal, Khira-90, Puna khira-1, Pant khira, Swarn ageti, Swarn sheetal Swarn poorna	Amrit, Pusa sanjog, Pant sankar khira-1, Hybrid-1, Noori, rani, Sangeeta, Mathura, Kasinda, Kiros
Pumpkin	Pusa viswas, Kashi harit, Narendra agrim, Narendra amrit, Arka chandan, Arka suryamukhi, Ambili	Sonia, Arjun, Chipo
Summer squash	Punjab chhappan kaddu-1, Pattyapan, Australian green	Pusa alankar
Ash gourd	CO-1, CO-2, IVAG-90, Kashi dhawal, Indu, Barailly local, IVAG-502	-
Snapmelon	Local varieties	-
Snake gourd	CO-1, CO-2, Kamuadi, PKM-1, Baby	-

Source: Rai *et al.*, 2004 and 2006.

Climate

Climate of diara/river basin is favourable for the cultivation of cucurbitaceous crops. Climate of such areas is not very much variable and seems almost same all round the year. High day temperature and cool nights are the main feature which makes is favourable for early cultivation of cucurbits. In diara land areas major part of soil is sand which warms up easily during day time to provide favourable temperature for seed germination. In these areas untimely rain also does not adversely affect the crop because sandy soils have less water holding capacity which also leads better growth and development of plants.

Soil and its Preparation

River basin areas are formed and subject to alluvium and diluvium action of perennial Himalayan rivers and due to inundation caused swollen rivers during South-West Monsoon. Fresh silt and clay deposits received every year, during the monsoon months, especially in Himalayan rivers, makes these land suitable for growing vegetable crop, literally on sand. Even though upper layers of land seem unsuitable for growing crops, the subterranean moisture seeped from adjacent river streams, makes it possible to grow early crop (Seshadri, 1986). For land preparation generally there is not need of ploughing at all. Farmers only have to uproot weeds and other unwanted wild species manually from soil surface.

Preparation of Trenches/Channels/Pits

After the cessation of South-West monsoon and recession of flood during October-November, river beds are ready to prepare channels/trenches and digging pits in the month of November. The channels/trenches/pits should be dug in North-South direction in lengthwise so that trenches are open east direction to manage the availability of temperature and moisture. The trenches/channels should be kept 45-60 cm wide and 45 to 90 cm deep (depending upon the height of water table because generally 60-90 cm is the height of water table in the river basin areas). Length of channels/trenches is variable and depends upon area and location. Circular and semi-circular pits of about 30-45 cm in diameter having depth of about 90 cm can be prepared for sowing the seeds of cucurbits. These channels/trenches or pits should be left as such for few days and if possible pre-sowing irrigation should be provided to settle down the various layers of materials in trenches (Pandey *et al.;* 1979, Yadav, 2003 and De *et al.,* 2003).

Nutrient Management

In the past, farmers were not applying manures and fertilizers but now it is observed that farmers are paying attention towards nutrient management. It is not standardized but on the basis of observations, it is recommended that 4-5 kg well rotten FYM/compost, 100-150 g Castor/groundnut/Mandua cakes, 40 g urea, 80 g SSP, 60 g MOP and 3 g furadon should be applied per pit as basal dressing to get better production. To increase the water holding capacity in root zone silt should also be added in the mixture. At the time of thinning, application of area @ 30-60 g per pit is also useful for better growth and development of plants. 30 to 40 days after seed sowing 40 g urea again can be used as band placement. Before flowering one foliar spray of urea is also found useful.

Seed Rate, Spacing and Dowing

In general, seed rate varies according to crop, variety and planting distance, however, in river bed cultivation comparatively less seed is required for cultivation. For early crop, seed should be sowing in 1st or 2nd week of November and in some areas in the 1st week of December. Late sowing is done in the first week of January. Distance between trenches/channels/pits should be kept 2-3 m depending upon the crop and varieties to be raised while 50-90 cm between plant to plant should be maintained and seed should be sown at 3 to 4 cm depth. For better and early germination pre-sprouted seeds should be sown, when the temperature is very low. For the pre sprouting seeds should be kept moist for 24 hrs. and after that wrap it in a gunny bags and covered with cotton cloths and left for 2-3 days at warm place to get sprouted. As soon as sprouts are visible out side the seed coat, seeds should be sown in the pits or trenches. Generally 2-3 presoaked seeds should be used for sowing at one place.

Protection of Crops from Cold and Hot Winds

In North and north western India, the cucurbits when sown in the month of November, are able to with stand the cool temperature of winter months from December to February which normally goes down upto 1-2°C and in summer from April to June when temperatures upto 46-48°C, then there is need to provide some protection from such adverse conditions to crop in the form of thatch screen of one meter height which act as wind breaks, generally made up of locally available materials like *Saccharum* grass (kans/Sarkanda), *Ipomoea cornia* (Sadabahar), paddy straw, sugarcane leaves *etc.* It is helpful in providing sufficient protection on the dug up trenches and covering the hills sown with seeds and is available for spreading over sand when vines grow and cover the sand to stop the drifting of sand (Mangal and Singh, 1993 and Ram *et al.*, 2007).

Water Management

In river bed cultivation flood irrigation or irrigation through channels is not quite possible. In early stages, irrigation is given though pitcher (earthen pots). Uses of drip or sprinkler systems are beneficial to increase the water use efficiency, however, the cost of cultivation increase. During later stages, crop is rarely irrigated because cucurbits have long tap root system which enables them to get moisture from underneath the sand in river bed cultivation.

Weed Management

In diara land areas most of the weeds come through water and spread all the area when flood occurs during monsoon period. Major weed species of river bed areas are *Polygonum sp. Eclipta prostrate, Euphorbia hitra, Portulaca quardifida, Lathyrus sativa, Melilotus indica, Cyperus rotundus, Dicotoma* spp. *etc.* In these areas, weeds should be only controlled manually or by mechanical means. Hand weeding is a common and essential practice for the control of weeds in river bed areas. No use should be strictly pointed as it may lead to mix residues in running water of river and may prove hazardous to human, animals and fishes and other water bodies (De *et al.*, 2003).

Harvesting and Yield

In general, river bed crop yield more in comparison to normal crop and harvesting of crop starts from the month of February-March, however, the yield of different cucurbitaceous crop varies according to soil conditions, location and mixed cropping, seed quality and adoption of improved agronomical practices bed suited to river bed cultivation. The harvesting time and yield of various cucurbitaceous crops under riverbed cultivation are given in Table 18.3.

Table 18.3: Harvesting Time and Yield of Cucurbits under River Bed Cultivation

Crop	*Harvesting*	*Average Yield (q/ha)*
Bottle gourd	March-July	380-450
Bitter gourd	May-July	300-360
Ridge gourd	April-July	230-290
Sponge gourd	April-July	280-325
Muskmelon	March-June	200-325
Watermelon	March-July	410-450
Longmelon	Feb.-March	250-280
Pointed gourd	March-July	300-400
Cucumber	March-June	250-310
Pumpkin	March-June	480-530
Summer squash	April-July	150-175
Ash gourd	April-June	200-350
Snapmelon	April-June	225-250
Snake gourd	March-June	150-175

Major Constraints

- ☆ Farmers utilizing river beds for vegetable cultivation, are resource poor and illiterate, therefore, they are not utilizing bank credit facilities as provided by government.
- ☆ Breeders not paying their attention to bred improved/hybrid varieties best suited to river bed areas.
- ☆ There is not researches at all on agronomic aspects *viz.* seed rate, spacing, fertilizer management especially for river bed cultivation.
- ☆ Infestation of fungal diseases like downey mildew, powdery mildew, anthracnose and fusarium wilt and viral diseases like cucumber green mosaic virus seed born musk melon mosaic virus, melon mild mosaic virus and ring spot virus are more prevalent. Insects like red pumpkin beetle and aphids are usually noticed in early stages while fruit fly and mites are more prevalent in later stage.
- ☆ Lack of awareness among the growers about hybrid crop production techniques.

- ☆ Lack of silt and rich subsoil beneath the sandy layer and leaching of nutrients due to sandy substrata cause deficiencies of major and minor nutrients.
- ☆ Cucurbits are cross pollinated in nature because of that crop produced in this system are not uniform and of standard quality especially in the sweetness of melon fruits.

Research Priority Areas

- ☆ There is need to develop improved/hybrid and multiple resistant varieties especially for river bed areas.
- ☆ There is need to develop crop and location specific effective production techniques for improved/hybrid varieties and while developing these, poor resources of small and marginal farmers should be taken into account.
- ☆ Development of sound post harvest management techniques and established network of cool chains and cold rooms will be a boon for the vegetable growers of river bed areas.
- ☆ Promotion of cooperative society will ensure adequate supply of quality seed as well as other inputs at proper time.

References

De Nirmal, Pandey Sudhakar, Singh, K.P. and Rai, M. 2003. Integrated development of gourds melons, IIVR, Technical Bulletin 13, pp. 28-33 Varansi, India.

Mangal, J.L. and Singh, G.R. 1993. Off season vegetable production. In Advance in Horticulture vol. 6. Vegetable Crops : Part 2 (Chadha K.L. and Kalloo G. Eds.) Malhotra Publishing house, New Delhi, pp. 673-686.

Pandey, U.C. Lal, S. and Mangal, J.L. 1979. Nadi Kinare Sabji Ki Kheti. *Farmers and Parliament,* 14 : 35-37.

Rai, M., Singh Major, Pandey Sudhakar, Pandey, K.K., Singh Jagdish, Kumar, Sanjeet and Singh, B. 2006. IIVR : A decade of Accomplishments, IIVR pp. 6-31. Varanasi, India.

Rai, M., Kumar, S. Pandey, S. Singh, M. and Singh, B. 2004. Popular varieties of vegetable crops in India, Technical bulletin no.-27, IIVR Varanasi pp. 1-93.

Ram, D., Pandey, A.K., Rai, M., Pandey, Sudhakar and Rai, N. 2007. Diara Kchhetra Me arthik Samridhi Ke Liya Kadduvargiya Sabjiyan. Smarika Regional Agricultural Fair held at IIVR from 3-5 Nov. 2007. Organized by IIVR, Varanasi Agricultural Cooperative Deptt. New Delhi and NHB Gurgaon Haryana, pp. 22-26.

Seshadri, V.S. 1986. Cucurbits, In vegetable crops in India (Bos T.K. and Som M.G. Eds.) Naya Prokash, Calcutta-6, pp. 91-164.

Seshadri, V.S., Chatterjee, S.S. 2000. The history and adoption of some introduced vegetable crops in India. *Asian Agri.* History, 4(3): 175-202.

Yadav, J.P. 2003. Nadiyon Ke Kinare Sabji Ki Kheti, *Kisan Jyoti.* 20(4) : 22-23.

Chapter 19

Testing of Hypothesis by using Excel and SPSS

Gaurav Shukla[1] *and Umesh Chandra*[2]

[1]*Department of Statistics,* [2]*Department of Computer Science, CoA, Banda University of Agriculture and Technology, Banda, U.P.*

1. Introduction

In many field investigations, an investigator is often interested in comparing some characteristic (such as mean or variance) of the population under study (group) with a specified value, or in comparing two or more groups with regard to some characteristic. For example, one may wish to know whether average yield of a crop in a certain district is equal to a specified value, one may wish to compare two varieties of a crop with regard to mean yield, to know if genetic fraction of total variation in a strain is more than a given value *etc.* Testing of hypothesis is one of the aspects of statistical inference, in which we draw conclusions about a population on the basis of a sample selected from it. Here, we make a statement about a population (the statement is usually called hypothesis) and on the basis of a sample drawn from it decides whether to accept or reject the statement.

Procedure of Testing Hypothesis

1. Set up a hypothesis
2. Set up a suitable significance level
3. Set a test criterion according to null hypothesis
4. Calculate the value of test statistic
5. Making decisions

2. Definitions

Statistical hypothesis: A statistical hypothesis is a conjecture or assumption about a population parameter. This conjecture may or may not be true. For example, average height of the students of a class is 165 cm.

If all the parameters are completely specified, hypothesis is called a simple hypothesis; otherwise it is a composite hypothesis. For example, average height of the students of a class is 165 cm is a simple hypothesis and average height of the students of a class is greater than 165 cm is a composite hypothesis.

Null Hypothesis

According to Prof. RA. Fisher, A null hypothesis is the hypothesis which is tested for its possible rejection under the assumption that it is true. It is denoted by H_0.

Alternative Hypothesis

Any hypothesis which is complementary to the null hypothesis is called an alternative hypothesis, usually denoted by H_1. For example, if we want to test the null hypothesis that the population has a specified mean μ_o. *i.e.* $\mu = \mu_o$, then alternative hypothesis would be

$\mu > \mu_o$ or $\mu < \mu_o$ or $\mu \neq \mu_o$

Two Types of Error: The main objective in sampling theory is to draw valid inferences about the population parameters on the basis, of the sample results. In the hypothesis-testing situation, there are four possible outcomes. In reality, the null hypothesis may or may not be true, and a decision is made to reject or not reject it on the basis of the result obtained from a sample. The four possible outcomes are shown in Figure. Notice that there are two possibilities for a correct decision and two possibilities for an incorrect decision.

	Ho is True	Ho is False
Reject Ho	Type I Error	Correct Decision
Accept Ho	Correct Decision	Type II Error

In practice we decide to accept or reject the lot after examining a sample from it. As such we are liable to commit the following two types of errors:

Type I Error: Reject Ho when it is true.

Type II Error: Accept Ho when it is wrong, *i.e.*, accept Ho when H_I is true.

Probability of type I error = α

Probability of type II error = β

Statistical Significance

When we examine a sample drawn from a population under study for testing the null hypothesis, the sample estimate of a certain characteristic is found differ from the population parameter. This difference can be attributed due to two causes:

1. It is arises due to errors of sampling
2. It is arises due to fact that the parent population of the sample is differ from the population considered.

If the difference between the sample estimate and the population parameter under the null hypothesis is due to error of sampling, it will have some limiting value. This limiting value (say δ) can be positive or negative. If the observed difference between the sample estimate and the population parameter is less than δ then we can say it is only due to sampling error alone and we accept our null hypothesis. And if this difference is $\geq \delta$, then we can say it is not only due to sampling error but also due to fact that the parent population of the sample is differ from the population considered. Hence we reject our null hypothesis.

Critical Region

A region (corresponding to a test statistic) in the sample space S, which amounts to rejection of null hypothesis is termed as critical region or region of rejection. The selection of critical region is based on the probabilities of two types of error. The probability of I type is fixed and the critical region is chosen which minimize type II error.

Level of Significance

The probability 'α' that a random value of the test statistic belongs to the critical region is known as the level of significance. In other word, we can say the level of significance is the maximum probability of committing a type I error or level of significance is the size of the type I error.

Critical Values or Significant Values

The value of test statistic which separates the critical or rejection region and the acceptance region is called the critical value or significant value. It depends upon:

1. The level of significance used, and
2. The alternative hypothesis, whether it is two-tailed or single-tailed.

One Tailed Test

A one-tailed test indicates that the null hypothesis should be rejected when the test value is in the critical region on one side of the sampling distribution. A one-tailed test is either a right tailed test or left-tailed test, depending on the direction of the inequality of the alternative hypothesis.

Two Tailed Test

If the direction of the inequality of the alternative hypothesis is not known, the critical region will be on either end of the sampling distribution, therefore the test is known as two tailed.

Degree of Freedom

Degree of freedom is the numbers of observations that are free to vary after certain restriction have been placed on data. If there are n observations in a sample, for each restriction imposed upon the original observations the number of degrees of freedom is reduced by one.

P-Value

The P-value or calculated probability is the probability of finding the observed or more extreme results when the null hypothesis is true. The definition of extreme is depends upon the condition, how the hypothesis being tested.

3. Test of Significance for Small Samples

Assumption for "t" Test

1. Sample should be drawn from a normal population.
2. Sample size should be smaller than 30.
3. Variance of the population should be unknown.
4. Sample should be drawn independently.

Test of Significance of Single Mean

Let $x_1, x_2, \ldots\ldots, x_n$ be a random sample from a normal population with mean μ and unknown variance σ^2. Samples are drawn independently.

Testing Hypothesis

We want to test a null hypothesis

$H_0: \mu = \mu_0$

Against $H_1: \mu > \mu_0$ or $\mu < \mu_0$ or $\mu \neq \mu_0$

Test Statistic

We use a test statistic

$$t = \frac{\bar{x} - \mu_0}{s/\sqrt{n}} \quad \sim t_{n-1}$$

$$\textit{Where} \quad s^2 = \frac{1}{n-1}\sum_i (x_i - \bar{x})^2 \quad \textit{and} \quad \bar{x} = \frac{1}{n}\sum_i x_i$$

We reject our null hypothesis at α per cent level of significance if

For $H_1: \mu > \mu_0$	$t > t_{\alpha:\, n-1}$	Otherwise accept H_0
For $H_1: \mu < \mu_0$	$t < t_{1-\alpha:\, n-1}$	Otherwise accept H_0
For $H_1: \mu \neq \mu_0$	$\lvert t \rvert > t_{\frac{\alpha}{2}:n-1}$	Otherwise accept H_0

SPSS Procedure for Test of Significance of Single Mean

1. To run a One Sample t Test in SPSS, after fitting the data in Variable view/ data window, click **Analyze > Compare Means > One-Sample T Test**.
2. Move variable (whose mean will be compared to the hypothesized population mean) to the **Test Variable(s)** area by clicking the **arrow** button. Write hypothetical population mean value in **Test Value Box**.
3. Clicking **Options** will open a window where you can specify the **Confidence Interval Percentage** then Click **Continue** and Click **OK** to run the One Sample *t* Test.
4. Now you will get **One-Sample Statistics Table** and **One-Sample Test Table.**

Test of Significance of the Difference between Two Sample Nean

Let $x_{11}, x_{12}, ...x_{1n_1}$ and $x_{21}, x_{22}, ...x_{2n_2}$ be two independent random samples from normal populations with mean μ_1, μ_2 and unknown variance σ_1^2, σ_2^2 respectively. It is assumed that both the variances are equal.

Testing Hypothesis

We want to test a null hypothesis

$H_0: \mu_1 = \mu_2$

Against $H_1: \mu_1 > \mu_2$ or $\mu_1 < \mu_2$ or $\mu_1 \neq \mu_2$

Test Statistic

We use a test statistic

$$t = \frac{\overline{x}_1 - \overline{x}_2}{s\sqrt{\frac{1}{n_1} + \frac{1}{n_2}}} \sim t_{n_1+n_2-2}$$

Where $$s^2 = \frac{\sum_i (x_{1i} - \overline{x}_1)^2 + \sum_i (x_{2i} - \overline{x}_2)^2}{n_1 + n_2 - 2} = \frac{(n_1 - 1)s_1^2 + (n_2 - 1)s_2^2}{n_1 + n_2 - 2}$$

We reject our null hypothesis at α per cent level of significance if

For $H_1: \mu_1 > \mu_2$	$t > t_{\alpha:n_1+n_2-2}$	Otherwise accept H_0
For $H_1: \mu_1 < \mu_2$	$t < t_{1-\alpha:n_1+n_2-2}$	Otherwise accept H_0
For $H_1: \mu_1 \neq \mu_2$	$\lvert t \rvert > t_{\alpha/2:n_1+n_2-2}$	Otherwise accept H_0

SPSS Procedure for Test of Significance of the difference between Two Sample Mean

1. To run an Independent Samples t-test in SPSS, after fitting the data in Variable view/data window, click **Analyze > Compare Means > Independent-Samples T Test** then we find **Independent-Samples T Test** dialogue box.
2. In dialogue box choose dependent variable in to **Test Variable Box** and transfer the independent variable in to **Grouping Variable Box** by using **Arrow** button.
3. Now click on Define Group and **Define Groups** dialogue box will present. In Define Group dialogue box enter 1 in to **Group 1** and 2 in to **Group 2** then Click the **Continue** button.
4. If you need to change the confidence level limits or changes how to exclude cases, click the **Option** button. Click the **Continue** button. You will be returned to the **Independent-Samples T Test** dialogue box. Click the **OK** button.
5. Now you will get Group Statistics Table and Independent Samples Test Table.

EXCEL Procedure for Test of Significance of the difference between Two Sample Mean

1. To run an Independent Samples t-test in EXCEL, after fitting the data in excel sheet, click **Formulae > More Function > Statistical > T Test** then we find **T Test** dialogue box.
2. Select first data set and put array in **Array 1** similarly select second data set and put array in **Array 2**, Write 1 in **tails** if one tail test is used or 2 if two tail test is used.
3. Write 2 in **Type** for two sample t test with equal variance or 3 for two sample t test for unequal variance.
4. Then click **OK.**

Test of Significance of the difference between Two Sample Means when the Samples are Correlated (Paired t test)

Let $(x_1, y_1), (x_2, y_2), \ldots\ldots\ldots (x_n, y_n)$ be n pair of observations (samples) from a bivariate normal population whose all parameters are unknown and d be the difference between the corresponding pair.

Testing Hypothesis

We want to test a null hypothesis

$\mathbf{H_0}: \mu_d = 0$

Against $\mathbf{H_1}: \mu_d > 0$ or $\mu_d < 0$ or $\mu_d \neq 0$

Test Statistic

We use a test statistic

$$t = \frac{\bar{d}}{s_d / \sqrt{n}} \qquad \sim t_{n-1}$$

Where $s_d^{\,2} = \frac{1}{n-1}\sum_i \left(d_i - \bar{d}\right)^2$ *and* $\bar{d} = \frac{1}{n}\sum_i d_i$

We reject our null hypothesis at α per cent level of significance if

For $\mathbf{H_1}$: $\mu_d > 0$	$t > t_{\alpha;\, n-1}$	Otherwise accept H_0
For $\mathbf{H_1}$: $\mu_d < 0$	$t < t_{1-\alpha;\, n-1}$	Otherwise accept H_0
For $\mathbf{H_1}$: $\mu_d \neq 0$	$\lvert t\rvert > t_{\frac{\alpha}{2};n-1}$	Otherwise accept H_0

SPSS Procedure for Paired t-test

1. To run a Paired Samples t-test in SPSS, after fitting the data in Variable view/data window, click **Analyze** > **Compare Means** > **Paired-Samples T Test** then we find **Paired-Samples T Test** dialogue box.
2. Write 1 on **pair** slot in the **Paired Variables box** and Double-click on a variable to move it to the **Variable1** slot in the **Paired Variables box**. Then double-click on another variable to move it to the **Variable2** slot in the **Paired Variables box**.
3. Clicking **Options** will open a window where you can specify the **Confidence Interval Percentage** then Click **Continue** and Click **OK** to run paired t test.
4. Now you will get **Paired Sample Statistics Table, Paired samples correlation** and **Paired Samples Test Table.**

EXCEL Procedure for Test of Significance of the difference between Two Sample Mean

1. To run an Independent Samples t-test in EXCEL, after fitting the data in excel sheet, click **Formulae > More Function > Statistical > T Test** then we find **T Test** dialogue box.
2. Select first data set and put array in **Array 1** similarly select second data set and put array in **Array 2**, Write 1 in **tails** if one tail test is used or 2 if two tail test is used.
3. Write 1 in **Type** for paired t test
4. Then click **OK.**

Test of Significance of Correlation Coefficient

Let $(x_1, y_1), (x_2, y_2), \ldots\ldots\ldots (x_n, y_n)$ be n pair of observations (samples) from a bivariate normal population whose all parameters are unknown

Testing Hypothesis

We want to test a null hypothesis

$H_0: \rho = 0$

Against $\quad H_1: \rho \neq 0$

Test Statistic

We use a test statistic

$$t = \frac{r\sqrt{n-2}}{\sqrt{1-r^2}} \quad \sim t_{n-2}$$

$$\text{Where} \quad r = \frac{\frac{1}{n}\sum xy - (\overline{x}\,\overline{y})}{\sqrt{\left(\frac{1}{n}\sum x^2 - (\overline{x}^2)\right)\left(\frac{1}{n}\sum y^2 - (\overline{y}^2)\right)}}$$

We reject our H_0 at α per cent level of significance if $|t| > t_{\alpha/2:n-2}$, otherwise accept H_0.

SPSS Procedure for Test of Significance of Correlation Coefficient

1. To run a bivariate Pearson Correlation in SPSS, after fitting the data in Variable view/data window, click **Analyze > Correlate > Bivariate.**
2. Select the variables listed in the left and click the blue **arrow** button to move them to the right, in the **Variables** field. There are multiple types of correlation coefficients in **Correlation Coefficient Box**. Tick on **Pearson** for a bivariate Pearson Correlation.
3. Click **Two-tailed** or **One-tailed**, depending on your desired significance test in **Test of Significance Box** then **Tick** on **Flag Significant Correlation Box.**
4. Clicking **Options** will produce the options of **Means and standard deviations, Cross-product deviations and covariances** of the variables.
5. Click **OK** to run the bivariate Pearson Correlation.
6. Now you will get **Correlations Table.**

References

Agarwal, B.L. (2003). Elementary Statistics, New age international publishers, New Delhi.

Chandel, S.P.S. (1997). A handbook of agricultural statistics, Achal prakashan mandir, Kanpur.

Gomez, K.A. and Gomez, A.A. (1984). Statistical Procedures for agricultural research, John willy and sons, New York.

Goon, A.M., Gupta, M.K. and Dasgupta, B. (2005). Fundamentals of statistics Vol.-1, World Press, Culcutta.

Goon, A.M., Gupta, M.K. and Dasgupta, B. (2005). Fundamentals of statistics Vol.-2, World Press, Culcutta.

Gupta, S.C. and Kapoor, V.K. (2002). Fundamentals of mathematical statistics, Sultan chand and sons, New Delhi.

Hines, W.W., Montgomery, D.C., Goldsman, D.M. and Borror, C.M. (2003): Probability and Statistics in Engineering, John Wiley and Sons, New York.

Rangaswami, R. (2010). A text book of agricultural statistics, New age international publishers, New Delhi.

Tailor, R. K. (2011). Advanced Statistical Methods, Paradise Publisher, Jaipur.

Shukla, R. K. and Kumar, K. (2011). A text book of pharmaceutical biostatistics, Vigyan Bodh Prakashan, Agra.

Chapter 20

Ensuring the Enhancement of Farm Income through Suitable Price Policy and Support System

Abhishek Kalia[1], Snehal Mishra[2], Bhanu P. Mishra[3], B.K. Gupta[4], Dheeraj Mishra[4] and Pankaj K. Ojha[4]

[1]Assistant Professor, Department of Agricultural Economics
[2]Assistant Professor, College of Agribusiness Management,
Anand Agriculture University, Anand, Gujarat
[3]Associate Professor and Head, Department of Agricultural Extension
[4]Assistant Professor, Department of Agricultural Extension
Banda University of Agriculture and Technology, Banda, U.P.

Agricultural price policy in India was introduced since independence. But the agricultural price policy formulated in India has varied widely for different years and also for different crops. This policy put much emphasis on the prices of food grains like wheat, rice and coarse cereals such as jowar, bajra, maize *etc.* In India, the price policy was first introduced in 1947 with the formation of Food Grains Policy Committee which recommended a policy of progressive decontrol, reduction of imports or food grains and substantial increase in the production of foodgrains. Again in 1950, Foodgrains Procurement Committee was appointed which introduced the system of rationing and control in the supply of foodgrains in the country. The main objective of the price policy in India was to protect the interests of consumers. In this policy no attention was paid to provide incentive price to farmers. It was only in 1964, a clear-cut policy was introduced for providing incentive price to farmers. At present, the government decides the Administered Prices for various agricultural commodities taking into account the recommendations of the Commission for Agricultural Costs and Prices (CACP), the views of state governments and central ministries as well as such other relevant factors which are considered important for fixation of support prices for agricultural commodities. The administered prices are

Minimum Support Price

This is the price fixed by the Government to protect the producer-farmers against excessive fall in price during bumper production years. It is announced before the sowing season.

Procurement Price

This is the price at which the government procures the commodity from the producers/manufacturers for maintaining the buffer stock or Public Distribution System. It is announced before the harvesting season.

Issue Price

This is the price at which the government supplies the commodities at fair price shop and ration depots.

(The difference between Issue Price and MSP is called Subsidy)

MSP is determined by the Government by considering Cost of Production, Demand and Supply, Price trends in Domestic and International Price, Inter Crop Price Parity, Terms of trade in between Agricultural and Non agriculture sector. But even having such a sound price policy the availability of MSP to the farmers of the country is very low. In regard to this, Government of Madhya Pradesh in the year 2016-17 came up with Bhawantar Bhugtaan Yojana and Goernment of Orissa came up with a Direct Income/Investment Support Scheme named as KALIA (Krishuk Assurance For Livelihood Income Augmentation) The salient features of the Bhawantar Bhugtaan yojana are as follow:

1. Difference between the MSP and the Average Sale Price (ASP) of a produce in the APMC yard is paid directly to their bank accounts.
2. A total of **21.89 lakh** farmers with around **50 per cent** of total cultivable area was registered.
3. A major significant change in the procurement of Tur, Moong and Urad was noticed. (**480, 228 and 128 per cent** respectively)
4. Total cost of Procurement was also significantly reduced to **17.85 per cent.**
5. If market prices are higher, they can choose to sell in the open market and take advantage of high prices
6. And can receive the benefits of Bhavantar Scheme if market prices in later months are below MSP.

Besides, the Bhawantar Bhugtaan Yojana was prone to manipulation by traders and several mandi level functionaries. Since the time slot of registration and bringing produce to the market was very short and it depressed the market, with traders exploiting full advantage of it. Especially the small farms who generally sell their produce at farm level to aggregators, were the worst victims of this process. On the other hand, the DIS scheme like KALIA in orrisa and those running in Telangana and Karnataka to farmers on per hectare basis, are much simpler to implement, transparent, crop-neutral and without causing much distortion to the functioning of markets. The salient features of the scheme are as follows:

a. **Budget Allocated**- Rs. 10000 crores (2018-21)
b. **Government providing**- Rs. 5000 per acre for both seasons
c. **For landless farmers**- Rs. 12500 for allied activities
d. **Life and Accidental Insurance**- Rs. 2 Lakhs
e. **Best Part**- Rs. 10000 Annual assistances for those who are not able to do farming now due to old age, disability, disease or other reasons.

So this was all a small comparison among the three currently running schemes against the price distortion and ensuring the enhancement of the farm income through improved price policy support. Other than this, some other steps can be taken up by the government to improve the conditions of farmers are providing farm mechanization through Custom Hiring Centers, proper implementation of Pradhan Mantri Fasal Bima Yojana (PMFBY), providing local level storage, Creating Markets for Assured Returns to Farmers and by promoting Agro-Processing Units.

References

Commission for Agricultural Costs and Prices, Price Policy for Kharif Crops, The Marketing Season-2017-18, Commission for Agricultural Costs and Prices, 2018.

Gulati *et al.* (2018). Supporting Indian Farmers: Price Support or Direct Income/ Investment Support? (ICRIER).

Saini, S. and Gulati, A. (2017). Price Distortions in Indian Agriculture (ICRIER and The World Bank).

Chapter 21

Fruit Based Cropping Models for Enhancing Farmer's Income in Rainfed Bundelkhand of Uttar Pradesh

A.K. Srivastava[1], *Nidhika Thakur*[2] *and Om Prakash*[2]

[1]*Associate Professor and Head Department of Fruit science*
[2]*Assistant Professor Department of Fruit Science*
Banda University of Agriculture and Technology, Banda – 210 001, U.P.

Bundelkhand is very important region of India. It is unique in many aspects being the central part of the country. Bundelkhand is spread over Southern Uttar Pradesh and Northern Madhya Pradesh, between 23°10′ and 26°30′ north latitude and 78°20′ and 81°40′ east longitude. The region covers a geographical area of around 70,000 sq km and includes seven districts of Uttar Pradesh and six districts of Madhya Pradesh. These regions are characterized by low and erratic rainfall and harsh climatic conditions. Inspite of climate and other constraints, Bundelkhand region encompass many plant species which are being used as food, shelter, clothing and medicines by the people of village communities. Plants are the main source of socio-economic development as well as provide several things like food, fruits, flowers, fodder, fibre, fragrance, gum, resin, oil, spices, vegetable, dyes, rubber, wood, timber *etc.* Many fruit species like ber, bael, mahua, jackfruit, custard apple, pomegranate, aonla, datepalm and citrus *etc.* are grown commercially. Besides these fruit crops some underutilized fruits crops mostly indigenous ones are becoming popular in arid regions on account of their diversified uses. Underutilized crops could be considered in these areas for integration in existing farming system as these crops require less inputs and can sustain the harsh climatic conditions. There are around 30 plant species in arid zone known for their edible uses out of which some 20 plant species are known for their edible fruits or as vegetables *viz.*, kair, lasora, jharber, pilu, khejri, phalsa, bael, karonda, fig, dragon fruit, chirongi and prickly pear *etc.* These underutilized fruits have many advantages in terms of ease in growing, hardy nature and good yield even under extreme weather conditions. These crops have their own history of consumption because, local people are well aware about their nutritional and medicinal values. Thus, these crops have the potential to provide great nutritional and social benefits and it is increasingly advised that they are identified, researched and promoted in the way similar to the other commercial fruit crops.

Bundelkhand is an old landmass composed of horizontal rock beds, undulating terrain with low rocky outcrops, narrow valleys, and plains. Prevailing soil types are a mixture of black and red, gravely and shallow in depth, and unable to retain moisture well. The region, in general, suffers from acute ecological degradation due to deforestation and consequent top soil erosion leading to low productivity of the land. Soil erosion is a persistent problem that is aggravated because of the hilly terrain. The widespread growth of gullies is a common phenomenon. The region is marked by extremes of temperature reaching 48°C during the summer months and dropping sometimes as low as 1°C in the winter. The temperature begins to rise in February and reaches its peak in May-June.The rainfall distribution pattern is irregular, with approximately 90 per cent of total rainfall in the region caused by the monsoon falling from June to October. Average rainfall per year is 800-900 mm 85 per cent of it is lost as run-off. The Bundelkhand region was densely forested till the late 18th century. After the turn of the century, rising demands for wood and other forest produces, population explosion, thrust on developmental activities and agricultural expansion led to increasing levels of deforestation.

Table 21.1: Classification of Rainfall

Rainfall (mm)	*Zone*	*Net Sown Area*	*Rainfall*
< 500	Arid	16	Very low
500-750	Semi Arid	17	Low
750-1100	Dry Sub humid	35	Medium
1100-1400	Moist Sub humid	24	High
> 1400	Humid mountains	8	Very High

Sourse: Ramkrishna 1997; Paul, 1997.

Table 21.2: Delineating Criteria for Dryland vs. Rainfed Farming

Constituent	*Dryland Farming*	*Rainfed Farming*
Rainfall(mm)	< 800	> 800
Moisture availability	Shortage	Enough
Growing season (days)	< 200 days	< 200

Constituent	*Dryland Farming*	*Rainfed Farming*
Growing regions	Arid, Semi-arid, uplands of sub humid and humid regions	Humid and Sub humid regions
Cropping system	Single crop or intercropping	Intercropping or double cropping
Constraints	Wind and Water erosion	Water Erosion

Source: United Nations Economic and Social Commission for Asia and Pacific.

Cultivated area in the U.P. part of Bundelkhand is about 60 per cent and in the M.P. part about 42 per cent. Because of the uncertainty of water availability to the crops, people in the greater part of this region have to depend mostly on dryland farming. People's lives suffer both ways. During the rains, there is heavy runoff, carries off the topsoil, forms gullies, damages the properties as well as human cattle life. During summer most of the tanks, streams and lakes dry up.

Redeeming Features

In the northern part, the alluvial plains have great potential for groundwater recharge except where the base rocks exist at shallow depths. Much of these low land soil is black, highly retentive of rainwater. In the southern and south-western parts, there are considerable sandstone and limestone formations: these rocks having primary/secondary porosities are traditionally known as good absorbers of water, hence potential repositories of surface/ground water, even springs. Groundwater utilization is always costlier. Hence attention must primarily be given to tile conservation and augmentation of surface water resources. Conservation implies (i) minimizing the runoff to the river and (ii) preventing wrong uses. Earlier Bundelkhand was known for its dense forests (As mentioned in epic Ramayana). The present mining operations in Bundelkhand are so indiscriminate that the entire environment has been endangered. Mining starts with deforestation and leads to groundwater wastage, soil erosion, silting and pollution of rivers and streams. The prominent destroyed districts are Lalitpur, Jhansi, Panna, Mahoba, and Banda.

The greatest consequences in poor development of Bundelkhand were the destruction of forests and mining activities. It changed the climate, the rainfall pattern, the perennial character of the rivers, the groundwater recharge pattern, soil productivity, people's life pattern, social way of life and culture.

Table 21.3: Area, Production and Productivity of Horticultural Crops in Uttar Pradesh

Crops	*Area (Lakh ha)*	*Production (Lakh MT)*	*Productivity (MT/ha)*
Fruits	4.21	94.59	22.43
Vegetables	5.92	116.78	19.70
Spices	0.86	2.87	3.32
Total	17.42	356.36	

Directorate of Horticulture UP.

Table 21.4: Identification of Horticultural Plant Species for Dryland Areas

Eco-regions	*Plains*	*Plateaus and Sub-mountain Region*	*Temperate Hills*
Arid (less than 500 mm rainfall)	Ber (*Ziziphus mauritiana*)	Custard apple (*Annona squamosa*)	Chilgoza nut (*Pinus gerardiana*)
	Boardi (*Ziziphus rotundifolia*)	Bael (*Aegle marmelos*)	Seabuckthorn (*Hippophae rhamnoides*)
	Jharberi (*Ziziphus mauritiana*)	Ber (*Ziziphus mauritiana*)	Wild apricot (*Prunus americana*)
	Bael (*Aegle marmelos*)	Ker (*Capparis decidua*)	Chuli
	Lasoda (*Cordia myxa*)	Jharberi (*Ziziphus mauritiana*)	
	Aonla (*Emblica officinalis*)	Karonda (*Carrisa carandas*)	
	Date palm (*Phoenx dactylifera*)		
	Pilu (*Salvadora oleoides*)		
	Karonda (*Carrisa carandas*)		
	Mulburry (*Morus alba*)		
	Ker (*Capparis decidua*)		
	Phalsa (*Grewia subinaequalis*)		
	Drumstic (*Moringa oleifera*)		
Semi – Arid	Ber (*Ziziphus mauritiana*)	Aonla (*Emblica officinalis*)	Chestnut
	Jamun (*Syzygium cuminii*)	Bael (*Aegle marmelos*)	Mulberry (*Morus alba*)
	Tamarind (*Tamarindus indica*)	Custard apple (*Annona squamosa*)	Oleaster (*Elaeagnus angustifolia*)
	Wood apple (*Feronia limonia*)	Chirongi (*Buchanania lanzan*)	Himalyan crab apple (*Malus baccata*)
	Date palm (*Phoenx dactylifera*)	Wood apple (*Feronia limonia*)	Kafal (*Myrica esculenta*)
	Khirni (*Manilkara hexendra*)	Tendu (*Diospyros melanoxylon*)	Himalayan mulberry (*Morus serrata*)
	Palmayra palm (*Borassus flabellifer*)	Mahua (*Madhuca indica*)	Himalyan wild cherry (*Prunus cerasoides*)
	Mahua (*Madhuca indica*)	Monkey jack (*Artocarpus lakoocha*)	Mehal (*Pyrus pashia*)
	Tendu (*Diospyros melanoxylon*)	Karonda (*Carrisa carandas*)	Karu dhak (*Ribes glaciale*)
	Bellaric myrobalan (*Terminalia bellirica*)	Pomegranate (*Punica granatum*)	Himalayan yellow raspberry (*Rubus ellipticus*)
	Chebulic myrobalan (*Terminalia chebulic*)	Citron (*Citrus medica*)	Black berry (*Rubus fruticosus*)

Eco-regions	*Plains*	*Plateaus and Sub-mountain Region*	*Temperate Hills*
	Sweet lime (*Citrus limettioides*)	Indian butter tree (*Diploknema butyracea*)	Mauritius raspberry (*Rubus rosefolius*)
	Wild datepalm (*Phenix sylvestris*)	Jamoa (*Syzyzium fruiticosum*)	Kateru (*Prunus persica*)
	Indian almond (*Terminalia catappa*)	Indian almond (*Terminalia catappa*)	Amlok (*Diospyrous lotus*)
	Kichli (*Citrus maderaspatana*)	Galgal (*Citrus limon*)	Indian olive (*Olea ferruginea*)
	Nepal ebony persimmon (*Diospyros tomentosa*)	Attani (*Citrus rugulosa*)	Wild pomegranate (*Punica granatum*)
Sub humid	Wood apple (*Feronia limonia*)	Wood apple	Wild strawberry (*Fragaria nilgerrensis*)
	Khirni (*Manilkara hexendra*)	Jackfruit	Mysore raspberry (*Rubus niveeas*)
	Jackfruit (*Artocarpus hetrophyllus*)	Monkey jack	Sand pear (*Pyrus pyrifolia*)
	Jamun (*Syzizium cuminii*)	Indian persimmon	Mountain papaya (*Carrica candamarcensis*)
	Indian persimmon (*Diospyros peregrine*)	Nepal ebony persimmon	Jalpai (*Elaeocarpus floribiundus*)
	Ne pal ebony persimmon	Gamboge tree	
	Tendu	Egg tree	
	Gorgan nut (*Euryale ferox*)	Governer's plum	
	Gambose tree (*Garcinia cambogia*)	Wild banana (*Musa balbisiana*)	
	Egg tree (*Garcina xanthochymus*)		
	Governer's plum (*Flocartia ramontchi*)		
	Stargooseberry (*Phylanthus acidus*)		
	Wild datepalm		
	Aini (*Artocarpus hirsuta*)		
	Kokum (*Garcinia indica*)		
Humid prehumid	Citron (*Citrus indica*)	Monkey jack	Indian crab apple (*Docynia indica*)
	Wild mango (*Mangifera sylvatica*)	Amra	Mehal
	Wild banana (*Musa balbisiana*)	Citron	Amlok
	Amra (*Spondias mangifera*)	Wild banana	Sand pear
	Governer's plum	Governer's plum	Himalayan yellow raspberry

Eco-regions	*Plains*	*Plateaus and Sub-mountain Region*	*Temperate Hills*
	Puneala plum (*Flacourtia jangomas*)	Puneala plum	Mysore raspberry
	Star gooseberry	Latka	Puneala plum
	Jalpai		Kafal (*Malus sikkiminsis*)
	Jackfruit		Black berry
	Longan (*Euphoria longan*)		Mautitius raspberry
	Cowphal (*Garcinia cowa*)		Nilgiri strawberry

Table 21.5: Important Vegetables and Spices for Arid Regions

Vegetable crops for arid regions	Water melon or Mateera (*Citrullus lanatus*), round melon (*Citrullus fistulosus*), snap melon (*Cucumis melo*), Kachri (*Cucumis melo varagrestis*), Kakdi (*Cucumis species*), bottle guard (*Lageneria scicereria*), ridge guard (*Luffa acutangula*), bitter guard (*Momordica charantia*), clusture bean (*Cyamopsis tetragonoloba*), cow pea (*Vigna unguiculata*), Sem (*Lablab purpureus*), Okra (*Ablemoschus esculentum*)
Vegetable crops for irrigated arid land	Water melon, musk melon (*Cucumis melo*), bottle guard, ridge guard, sponge guard (*Luffa cylindrica*), bitter guard, snap melon, kachri, kakdi, tinda, onion (*Allium cepa*), garlic (*Allium sativum*), tomato (*Lycopersicum esculantum*), chilli (*Capsicum annum*), brinjal (*Solanum melongena*), okra, cabbage (*Brassica oleracea* var. *capitata*), cauliflower (*Brassica oleracia* var. *Botrytis*), Knol khol (*Brassica oleracea* var. *caulorapa*), pea (*Pisum sativum*), clusture bean, palak (*Beta vulgaris* and *Spinacia oleracea*) and leafy vegetables
Vegetables crops for spices	Cumin (*Cuminum cyminum*), fenugreek, funnel (*Foeniculum vulgare*), Coriender (*Coriendrum sativum*)

Table 21.6: Forest Trees, Shrubs and Pasture Grasses in Dryland Areas

Forest tree	Khejri (*Prosopis cineraria*), Vilayti babool (*Prosopis juliflora*), Kumat (*Acacia sengal*), Farash (*Tamarix articulata*), Rohida (*Tecomella undulata*), Babool (*Acacia tortilis*), Neem (*Azadirachta indica*), Mopane (*Colophospermum mopane*), Shisam (*Dalbergia sisoo*), Dhak (*Butea monosperma*)
Shrubs	Ker (*Caparis deciduas*), phog (*Calligonum polygonoides*), Aak (*Calotropis procera*), Euphorbia (*Euphorbia antisphylitia*), Kheep (*Laptodania pyrotechnica*), Aloe vera (*Aloe* spp.), Bhui (*Aerva tomentosa*)
Pasture grasses	Sevan (*Lasiurus sindicus*), Dhaman (*Centurus ciliaris*), *C. stigerus*, Karad (*Dicanthium annulatum*), Gorda (*Chrysopogon fulvus*)

Table 21.7: Fruit Growing Opportunity in Bundelkhand Zone of Uttar Pradesh

Fruit Crop	*Area*
Aonla	Jhansi, Lalitpur, Banda, Chitrakoot
Guava	Banda, Jhansi, Lalitpur Jhansi, Jalaun, Hamirpur, Mahoba, Chitrakoot, Banda, Lalitpur
Ber	Jhansi, Lalitpur, Hamirpur, Banda, Chitrakoot, Mahoba
Bael	Jhansi, Lalitpur, Hamirpur, Banda, Chitrakoot, Mahoba
Mandarin	Jhansi, Hamirpur, Lalitpur, Jalaun, Mahoba, Chitrakoot, Banda

Fruit Crop	*Area*
Sweet orange	Jhansi, Hamirpur, Lalitpur, Jalaun, Mahoba, Chitrakoot, Banda
Lim/Lemon	Jhansi, Hamirpur, Lalitpur, Jalaun, Mahoba, Chitrakoot, Banda
Pomegranate:	Jhansi

Silvi-Pasture and Horti-Pasture Systems

- *Acacia nilotica* + natural grass – For arid and semi-arid region
- *Acacia leucocephala* + natural grasses
- *Acacia catechu* + natural grasses
- *Dalbergia sisoo* + Barbar grasses
- *Prosopis cineraria* + Grasses
- *Zizyphus rotundrifolia* + Dhaman (*Cenchrus ciliaris*)
- *Zizyphus rotundrifolia* + Sewan (*Lasiurus sindicus*)
- Datepalm + natural grasses

Table 21.8: Fruit Based Models

Fruit +Fruit	Aonla + Ber +Karonda Ber + Guava + Pomegranate Bael + Ber + Phalsa Mulbery + Guava + Lime
Fruit + Vegetables	**Kharif** Bael + Cluster bean (Veg Type) Aonla + Kharif onion Ber + Amaranths (Grain type) Guava + Tuber crops **Rabi** Bael + Broad Bean Aonla + Coriander Ber + Tomato/Chilli/Brinjal Guava + Vegetable pea
Fruits + Agronomical Crops	**Kharif** Bael + Urd Aonla + Green Gram Ber + Sesame Guava + Cow pea (pulse type) **Rabi** Bael + Gram Aonla + Field Pea Ber + Lentil Guava + Linseed

Chapter 22

Recent Advances in Sugarcane Production Technologies

M.K. Tripathi[1], A.K. Chaubey[2], S.R. Singh[1], V.K. Singh[1] and A.P. Dwivedi[1]

[1]Principal Scientist (Agronomy), ICAR-IISR, Lucknow
[2]Associate Professor, Department of Soil Science, BUAT, Banda, U.P.

1. Introduction

Sugarcane is a most important cash crop of India. It involves less risk and farmers are assured up to some extent about return even in adverse condition. Sugarcane provides raw material for the second largest agro-based industry after textile. The sugar industry is an instrumental in generating the sizable employment in the rural sector directly and through its ancillary units. It is estimated that about 50 million farmers and their dependents are engaged in the cultivation of sugarcane and about 0.5 million skilled and unskilled workers are engaged in sugar factories and its allied industries. The sugar industry in India has been a focal point for socio-economic development in the rural areas by mobilizing rural resources, generating employment and enhancing farm income. There are 716 installed sugar factories (Co-operative-326, Private-347 and Public-43) in the country as on 31.01.2016, with sufficient crushing capacity to produce around 330 lakh MT of sugar.

The present day commercial sugarcane varieties are man-made hybrid clones involving *Saccharum officinarum* L., and *S. spontaneum* with a few genes incorporated from, *S. barberi*Jesw. *S. sinense* Roxb. and to a limited extent *S. robustum* Brandes. Prior to the development of the man-made varieties the *S. officinarum* clones were being cultivated in the tropical areas. Clones belonging to *S. barberi* and *S. sinense* were cultivated in northern India and parts of China in the sub-tropics.

Sugarcane is a tropical and subtropical, perennial grass that forms lateral shoots at the base to produce multiple stems, typically 3 to 4 meters high and about 5 cm in diameter. The stems grow into cane stalk, which when mature contributes about 75 per cent of the entire plant. A mature stalk is typically composed of 11 to 17 per cent of fibre, 12 to 16 per cent of soluble sugars, 2 to 3 per cent of non- sugars, and 60 to 70 per cent of water. As sugarcane crop is sensitive to the weather, soil type, irrigation, fertilizers, insects, disease control, cultivars, and the harvest period, care should be taken to follow best crop management practices for higher yield of sugar cane. Sugarcane belongs to bamboo family of plants and is indigenous to India. It is the main source of sugar. It also provides raw material for manufacturing alcohol. The part of sugarcane is also being used as fodder for animals. Sugarcane belongs to the grass family of "Poaceae" and genus of Saccharum".

2. Major Sugarcane Growing States

Sugarcane is grown in various states in subtropical and tropical regions of the country. Main sugarcane growing States are:

2.1 Sub-Tropical

Uttar Pradesh, Uttarakhand, Haryana, Punjab and Bihar with an annual rainfall of 180 to 2000 mm. The climate ranges from humid, moist sub-humid and dry sub-humid to cold arid, semiarid and arid.

2.2 Tropical Region

Karnataka, Tamil Nadu, Maharashtra, Andhra Pradesh, Gujarat, Madhya Pradesh with an annual rainfall of 602 to 3640 mm having moist to dry sub humid and semi-arid to dry semi-arid climates.

3. Important Regions/Zones for Sugarcane Cultivation in India

Broadly there are two distinct agro-climatic regions of sugarcane cultivation in India, *viz.*, tropical and subtropical. Tropical region shared about 45 per cent and 55 per cent of the total sugarcane area and production in the country, respectively. Sub-tropical region accounted for about 55 per cent and 45 per cent of total area and production of sugarcane, respectively.

3.1 Tropical Sugarcane Region

The tropical sugarcane region includes the states of Maharashtra, Andhra Pradesh, Tamil Nadu, Karnataka, Gujarat, Madhya Pradesh, Goa, Pondicherry and Kerala. The coastal areas of A.P. and Tamil Nadu have high sugarcane productivity. Floods, water logging and diseases such as red rot are the main problems. In the tropical region climatic conditions are more favourable for its growth. It is cultivated with better package of practices and higher irrigation levels. The growing season is long with more equitable and favourable conditions without serious weather extremes. Being a tropical country, the agro-climatic conditions of tropical India favour higher sugarcane and sugar yields. The tropical region contributes about 55 per cent to the total cane production in the country. The average cane yields of the major states of the region including Maharashtra, Tamil Nadu, Karnataka,

Andhra Pradesh and Gujarat is around 80 tonnes per hectare. Maharashtra and the adjoining area of Karnataka, Gujarat and A.P. record higher sugar recovery. Long hours of sunshine, cool nights with clear skies and the latitudinal position of this area are highly favourable for sugar accumulation. Moisture stress during the early part of the cane growths mostly during March to June is a major constraint in the state of Maharashtra and other part of region lacking perennial source of irrigation.

3.2 Sub-tropical Sugarcane Region

Around 55 per cent of total cane area in the country is in the sub-tropics. U.P, Bihar, Haryana and Punjab comes under this region. Extremes of climate are the characteristic feature of this region. During April to June, the weather is very hot and dry. July to October is rainy season accounting for most of the rainfall from South-West monsoon rains. December and January are the very cold month's temperature touching sub-zero levels in many places. November to March is cool months with clear sky. The cane yield is lower in the subtropics due to various reasons *viz.*, short growing season, high temperature disparity besides other factors like moisture stress, pest and disease problem, floods and water logging and very poor ratoon. The average yield of the four major states (U.P, Bihar, Punjab and Haryana) is around 60 tonnes per hectare. However, there is considerable potential to be exploited. Now the average yield of U.P. is around 71 tonnes/ha.

4. Varieties

Following are the important varieties but recommended varieties of the particular zone should be used:

(i) Early Varieties

CoS 8436, CoS 08279, CoS 13231, Co 0238, Co 0118, CoPk 05191, CoSe 01421, CoS 95255, CoS 96268, CoSe 95436, CoSe 98231, CoSe 00235, CoSe 01235, CoSe 03234, CoLk 94184 and CoP 9301 *etc.*

(ii) Medium Late Varieties

CoS 88216, CoS 91230, CoS 92263, CoS 96269, CoSe 98259, CoS 97264, CoS 96275, CoS 97261, CoS 94257, CoSe 92423, CoSe 95422, CoSe 96436, UP 39, CoSe 11453, Co 05011, CoS 09232, UP 9530, UP 0097, CoS 99259, CoS 07250, CoSe 01434, CoSe 01424, Co 0233, CoSe 08452 and CoPt 84212.

(iii) Late Varieties

CoS 94257, CoS 95255, CoSe 92423, CoSe 01424, CoLk 94184 and UP 0097.

(iv) For Limited Irrigation

CoLk 94184, CoS 96275, CoS 92263 and CoSe 92423.

5. Climate

Sugarcane can be grown as tropical and subtropical crop and grows well where temperature ranges between 20 C to 42°C. Sugarcane crop responds well to long period of sunlight, high humidity and sugarcane requires an average annual rainfall of between 1800 and 2500 mm for best crop growth. Sugarcane crop can be cultivated all along the year, if assured irrigation water is available. Sugarcane crop also withstands temperature variations of winter 6°C to 8°C and summer 40°C to 42°C.

Table 22.1: Temperature Requirement for different Growth Stages of Sugarcane

Critical Stages of Sugarcane	*Max. Temp. (°C)*	*Min. Temp. (°C)*	*Relative Humidity (Per cent)*
Germination	32.0	20.0	
Tillering	35.0	18.0	-
Grand growth	30.0	14.0	80.85
Ripening	30.0	20.0	50-55

6. Soil

Sugarcane crop can be grown in wide range of soils as it tolerates most of the soils as long as there is enough moisture content available in the soil. Heavy soils with good drainage are preferred for sugarcane cultivation, though it grows well on medium and light-textured soils also with assured irrigation. Soils with 0.5 to 0.6 per cent carbon content and pH 6.5 to 7.5 are most suitable for sugarcane growth. In northern India, it is cultivated largely on the loams and clay loams of Gangetic and other alluviums, and in peninsular India, it is grown on brown or reddish loams, laterites and black cotton soils. Sandy loam soils along the riversides are also good for sugarcane cultivation. As the sugarcane crop remains in the field over the year good fertile soils are most desirable. Waterlogged soils which have no drainage are not suitable for the cultivation of sugarcane. If the sugarcane is grown on large scale, it is advised to go for soil testing. Any soil nutrient deficiency should be compensated.

7. Land Preparation

Land preparation is another important task in sugarcane cultivation as this crop stands in the field for more than one year. With the help of tractor, land should be given 2 or 3 deep ploughings. First ploughing should be carried out immediately once the previous crop is harvested. Make sure to leave the land after ploughing for 2 to 3 months for sun expose. Any clods should be crushed and levelled. Second cross-wise ploughing should be carried by another plough and 3-4 harrowing should be given to bring the soil to fine tilth stage. Soil should be levelled with the help of planker for seed bed preparation. Supplement the soil with well-decomposed farmyard manure.

8. Time of Sowing and Harvesting

Sugarcane take generally one year to mature in sub-tropical states (U.P., Punjab, Haryana, Bihar *etc.*) called "Eksali" however in some tropical states it matures in 18 months (Andhra Pradesh, Karnataka, Maharashtra *etc.*) called "Adsali". In India, the planting seasons of sugarcane in different States is given below:

8.1 Autumn Planting

Planting of cane setts in October and November after harvesting of kharif crops. This season provides comparatively longer growing season (15 months) for germination of sugarcane thus higher cane yield compared to spring and summer seasons of planting. However, planting during this season is less preferred owing to the sacrifice of rabi crops. Farmers should accommodate one short duration crop as intercrop in between two rows of sugarcane. The suitable intercrops are wheat, oilseeds, pulses and vegetables.

8.2 Spring Planting

Most popular in the sub tropics. Planting of sugarcane is done after harvesting of short duration rabi crops like mustard, toria, potato *etc.* Planting is done during the month February and March. But the best time lies from Feb 15 to March 15.

8.3 Summer Season/Late Planting

Planting after harvesting of wheat crop in sub-tropical regions. Climatic conditions are harsh because of high temperature and desiccating atmosphere. Very short tillering span, low productivity, popular in Punjab, Haryana and western UP.

9. Planting Methods

9.1 Flat Planting

Land is prepared in a fine tilth by 1 or 2 deep ploughings followed by harrowing and planking. Shallow furrows of 10-15 cm depth at a distance of 90 cm in autumn and 75 cm in spring are opened and setts are sown followed by heavy wooden planking. Very common in north India. With the use of sugarcane cutter-planter developed by IISR it becomes very easy and economical.

9.2 Ring Pit Method

Also called as mother shoot technology or no tillage technology. Pits of 75cm diameter and 30 cm depth having centre to centre distance of 105 are dug. Total 9000 pits are made in one hectare. Suitable for drought prone areas, light textured soils, multi rationing, high yielding tall and thick varieties. This technique improves fertilizer use efficiency, weed control efficiency as well as water use efficiency.

9.3 Trench Method

Trenches, 30 cm wide and deep are made with centre to centre distance of 120 cm. Better WUE, better anchorage, crop lodging is less and higher yield under plant and multi ratooning too.

9.4 FIRB Method

Wheat is sown on the top of the beds and cane setts in the channels. Sowing 2-3 lines of wheat is done on beds in November and manual planting of sugarcane in furrows during spring season in the month of February and March.

9.5 Bud Chip Technology

Bud chips are scooped out from freshly harvested sugarcane stalk and planted in cup/trays filled with a mixture of soil, organic matter and sand in ratio of 1:1:1. The seedlings can be transplanted in the fields after 6-7 weeks.

9.6 Cane Node Technology

In this technology stem cutting having a single node/bud along with root band is taken and treated with liquid fungicide solution followed by treatment for 15-20 minutes in liquid organic formulation made up of cattle dung and cattle urine. Planting material is immersed in the above organic formulation. Then planting material is incubated in a nursery with adequate moisture. After five to six days buds start swelling and planted in the field at 30 cm distance with normal package of practices.

Paired Row System of Planting

- In the paired row system, two cane rows are brought together followed by a wide gap before the next set of two rows.
- The number of rows per hectare remains the same.
- The advantages are that wide spacing is available between the any two sets of paired rows which can be utilized for growing profitable intercrops. Due to good earthing up it is possible to prevent lodging. The system also permits better light interception by the crop and thus can give higher yields.

10. Seed Rate

Seed rate in sugarcane varies from region to region. Generally higher seed rate are used in north western India (Punjab, Haryana and Rajasthan) because of the lower germination per cent and also adverse climatic condition (very hot weather with desiccating winds) during tillering phase. A northern region seed rate generally varies from 40,000 to 60,000 three budded setts per hectares while in southern region it ranges from 25,000 to 40,000 three budded setts.

11. Selection of Planting Materials

- Healthy seed material free from pests and diseases like red rot, wilt, smut *etc.* should be selected for seed purpose.
- The top one-third to half portion of a cane being comparatively immature has buds of high viability and is best for sowing.
- If only upper half of the cane is utilized for planting, comparatively higher germination is secured.
- Seed cane should be taken from well manured, erect and healthy crop of not more than 8-10 months' age.
- Ratoon crop is not suitable for seed purpose as these canes may carry the disease of the previous crop.

- ☆ Priming of sugarcane setts with 50 ppm ethrel solution overnight before sowing enhances germination percentage.
- ☆ For best results separate crop nurseries should be raised specially for producing seed-canes under good crop management.
- ☆ Special precautions should be taken against disease and pest control.
- ☆ Optimum temperature for bud sprouting ranges from 20-30 degree celcius (not less than 10 and more than 40).

12. Water Management

In sugarcane, maintenance of optimum soil moisture during all stages of crop growth is one of the essential requisites for obtaining high yield. The crop should, therefore, be grown in areas of well-distributed rainfall or under assured and adequate irrigation. In tropical India, total water requirement of the crop for optimum growth varies from 2000 to 3000 mm inclusive of rainfall. The requirement of an adsali crop is proportionately higher (3200 to 3500 mm). In sub-tropical India, the water requirement is 1400-1800 mm. In tropical area, irrigations are to be given once in 7 days during germination phase (1 –35 days after planting), once in 10 days during tillering phase (36 – 100 days after planting), again once in 7 days during grand growth phase (101 – 270 days after planting) and once in 15 days during maturity phase (271 days after planting up to harvest) adjusting it to the rain fall pattern of the area. About 30 to 40 irrigations are needed. About 250 tonnes of water is needed to produce one tonne of sugarcane. Methods like alternate furrow irrigation, drip irrigation and trash mulching could be of use to economize irrigation water during water scarcity periods. Foliar spraying of a solution containing 2.5 per cent urea and 2.5 per cent Muriate of potash 3 or 4 times at fortnightly intervals during drought periods would help to reduce the impact of drought on the crop. Critical stages are those during which sugarcane is affected severely due to water stress and the loss cannot be compensated by adequate water supply at later stages. These stages are: sprouting (germination), formative stage or tillering, ripening and initiation of sprouting in ratoons. In case of limited water availability, one may sustain sugarcane productivity by irrigating at critical stages of growth.

13. Conservation of Irrigation Water

Even during periods of surplus irrigation water availability, flood irrigation has to be avoided and only controlled irrigation should be practiced so that surplus water could be conserved for use during deficit period. Furrow irrigation is the best method available to the farmer that is easy to adopt and wastage of the water can be minimised. Planting cane in long furrows and adoption of surge irrigation would be advantageous. Transporting of water from the source to the field head should be done as far as possible only through pipelines. At times of water scarcity, alternate furrow irrigation may be adopted. When water availability becomes favourable the farmer can switch back to normal furrow irrigation. In heavy soils, during germination phase, sprinkler irrigation would be advantageous to give light irrigation and to obtain better germination of the crop. Drip irrigation has the advantage of distributing the available quantity of irrigation water, however little it may be, to the entire crop area of a farm. The high cost of the system limits its adoption. Trash mulching can also be done wherever possible.

14. Fertilizer Requirement

A crop of 100 t/ha exhaust 208 kg N, 53 kg P and 280 kg K, besides 3.4 kg Fe, 1.2 kg Mn, 0.6 kg Zn, 0.2 kg Cu and 30 kg S. Indian soils are universally deficient in N, nearly 50 per cent soils are deficient in P and 20 per cent in K. Nutrient management practices have one of the major roles in crop growth, modification of juice composition and accumulation of sugar in cane stalk. An optimum N concentration of 1.95-2.0 per cent in plant at tillering has been estimated to be essential for maximum cane yield. The nitrogen requirement of sugarcane depends upon the soil and climate. It ranges from 150 kg/ha in Uttar Pradesh to 270 kg/ha in Tamil Nadu and 300 to 500 kg/ha in Maharashtra and Karnataka. Nitrogen is given in the form of urea applied one-third at planting and the remaining two-thirds in 2 equal splits at tillering and at the commencement of grand growth stage. The fertilizers may also be applied as basal dose through Diammonium phosphate to supply full P and part of N. The phosphorous is required at 40-60 kg of P_2O_5/ha. The response of sugarcane to potassium has been obtained only in localized pockets of light soils. Nowadays deficiency of sulphur is constantly increasing in Indian soils and it has become a limiting factor in sugarcane culture. In marginally deficient soils, the application of 40-60 kg S/ha has been found to be useful. Fertilizer recommendation for sugarcane in U.P. is 150:60:60:40 kg NPKS/ha in plant crop whereas in ratoon crop 25 per cent more nitrogen is advocated compared to plant crop.

14.1 Application of Bio-fertilizers

All soils contain substantial quantities of nutrients released by mineralization of organic matter, weathering, added through rain and irrigation water and biological fixation by native micro-flora. The soil contribution should be assessed through soil testing. Nutrient dosages are best decided by soil testing which would help accounting for the contribution of the soil. The balance nutrient requirement is to be met by chemical fertilizers, organic sources, bio-fertilizers, and green manures. The combination should be based on the availability of the source. At present, application of fertilizers is done without considering the contribution from other sources, even if they are applied. Though it is possible to grow sugarcane exclusively by organic means, it would be difficult to supply all the nutrient needs by organic means since obtaining huge quantities would be difficult and may be costly. Bio-fertilizers, supplemented by other sources can improve the soil status considerably. *Azospirillum* and phospho-bacteria are the two important and proven bio-fertilizers at present. *Gluconacetobacter* is yet another N fixing microorganism being extensively studied. *Azospirillum* application has been demonstrated to be highly beneficial particularly under low N soils and low N application rates and could supplement about 25-35 per cent N to sugarcane directly and can promote growth indirectly. phospho-bacteria has been found very effective in P solubilisation and can make P economy in the range of 25-50 per cent. Recently several strains of *Gluconacetobacter*have been identified and are being studied for their effectiveness. A suitable system could be to obtain about 50 to 75 per cent of the applied nutrients from fertilizers and the remaining from other sources (pressmud, green manure, bio-fertilizers *etc.*). A combination of chemical fertilizers + organic manure + bio-fertilizer has been found to be equal to growing sugarcane entirely through chemical fertilizers. There are many experimental results demonstrating the beneficial effects of organic sources.

14.2 Use of Sugarcane Trash

Sugarcane produces huge amount of foliage - about 40 per cent of the total biomass. About 15-20 t trash/ha is produced. Average nutrient content of trash is 0.35 per cent N, 0.13 per cent P_2O_5 and 0.15 per cent K_2O. For every tonne of cane trash nutrient availability is 3.5 kg N, 1.3 kg P_2O_5 and 6.5 kg K_2O. Thus for a hectare of sugarcane crop, around 50 kg N, 20 kg P_2O_5 and 100 kg K_2O can be recovered from trash alone. In sugarcane fields and in the vacant spaces of the farms, lot of weed growth occurs. Upon weeding, they are thrown away. Instead, all biomass could be used for composting and thus recycled. Trash could be used for mulching and then incorporation. It could be composted by various methods and sources. A novel technique is *in situ* composting particularly in ratoon and also upon removal by de-trashing. Sugarcane upon harvest leaves behind stubble and root mass to the extent of 4.5 per cent and 12.7 per cent of cane respectively. It is about 17.2 per cent of the total biomass produced and serves as a source of organic matter and nutrients. An estimate indicated about 4-5 tonne of dry matter contribution to a ratoon crop by the preceding plant crop. Its contribution in terms of nutrients works out to 50 kg N, 20 kg P_2O_5, 20 kg K_2O, 1.48 kg Na, 7.8 kg Ca, 8.12 kg Fe, 1.08 kg Mn, 0.07 kg Cu and 0.10 kg Zn per hectare.

14.3 Application of Green Manures

Green manure crops like Sunnhemp and Dhaincha have been a source of crop nutrients and soil productivity improvement. A preceding green manure crop always benefits succeeding sugarcane. Right time of sowing of green manure and ploughing in at right stage is important to realize the benefits. A good moisture status in the soil is essential for decomposition. After incorporation sufficient time should be available for taking up sugarcane planting. Green manure inter-cropping and *in situ* incorporation is an important practice in the integrated system. A preceding green manure crop could be given phosphatic fertilizer which could be utilized by the succeeding sugarcane.

15. Weed Management

- ☆ The slow germination and initial slow growth of the crop provides ample opportunity to weeds growth.
- ☆ The extent of crop loss due to the weeds varies from 10 to 100 per cent crop failure.
- ☆ Weeds can be controlled through cultural, mechanical and chemical methods.
- ☆ Critical crop weed competition period up to 90 days.
- ☆ Cultural –trash mulching and inter cropping.

- Intercropping with chickpea and peas in autumn and green gram, black gram and cow pea in spring planted sugarcane.
- Mechanical hoeing at 30, 60 and 90 day's stage of the crop.
- Chemical –Pre-emergence application of Atrazine @ 2kg a.i./ha Followed by 2,4-D spray @ 1.0 kg a.i./ha 60 days after planting.
- Sulfentrazone @ 0.5kg + atrazine @ 2.0 kg/ha pre – emergence proved very effective for controlling weeds in sugarcane specially *Cyperus rotundus*.
- For binding weeds (*Coccinia* spp., *Ipomoea* spp. and *Convolvulus arvensis*) in north Bihar and Eastern UP pre – emergence spray of Metribuzin @ 1.25 kg/ha.
- Integration of different methods of weed control provides better management than any single method.
- Otherwise operate small size power tiller fitted with for intercultivation.

16. Cropping Systems in Sugarcane

Crop production is harvesting of solar energy through crop plants. To maximize crop production in a given field, a major portion of the incident solar radiation has to be intercepted by an optimum leaf area throughout the year. This is not achieved in sole-cropping wherein the leaf area is sub-optimal in early growth and maturity phases. Besides this, in mono-cropping, that is, growing of the same crop year after year, usually there is long interval between the harvest of the crop in one year and planting of the crop in the subsequent year. During this long interval there is practically no leaf area to intercept the sunlight. Development and maintenance of an optimum leaf area throughout is feasible by combining crops in time and space utilizing the natural resources and applied inputs efficiently.

Sugarcane being a high biomass yielding long duration crop, exhausts large quantities of plant nutrients from the soil. Hence, raising of a good crop subsequently in the sugarcane field is not feasible without the addition of large quantities of plant nutrients through applied inputs like manures and fertilizers. When sugarcane is mono-cropped even with large quantities of applied inputs, as balance between removal and addition is not maintained in all nutrients, the soil gets depleted of some nutrients excessively. This leads to stunted growth of the crop, poor leaf area development and very low light interception and consequently the harvesting of solar energy is also very low. Combining crops with sugarcane in time and space helps in balanced use of both natural resources and applied inputs, facilitates development and maintenance of optimum leaf area throughout leading to maximum harvesting of solar energy and sustainable crop production.

Cultivation of sugarcane in a sequence with other crops can minimize all the ill-effects observed in mono-cropping of sugarcane. First of all there can be better utilization of all the natural resources of crop production and the applied inputs. As the different crops in the sequence are provided with varying ecological conditions, more balance is maintained in removal and addition of all plant nutrients. Any excess amount of nutrients applied to one crop is utilized by the next crop coming in sequence. Similarly any excess removal of nutrients by a crop is compensated by a subsequent crop receiving a heavy dose of fertilizers. As sequential cropping increases diversity, harmful organisms such as insects, diseases and problem weeds do not perpetuate. Because of these reasons, there is the possibility of normal growth of all the crops in sequence, which facilitates production and maintenance of optimum leaf area throughout leading to better interception and fixation of solar energy.

The available information indicates that for sustainable agriculture in areas where sugarcane is a dominant crop, either a 3-year or 4-year crop sequence should be adopted. Of these, two years will be occupied by one plant plus one ratoon crop of sugarcane while in the remaining one or two years compatible crops could be grown in sequence with sugarcane.

Table 22.2: Sugarcane Prevailing Cropping System in India

Cropping System for Sub-tropical	*Cropping System for Tropical Region*
Paddy- Autumn Sugarcane-ratoon-wheat	Bajra-Sugarcane(pre-seasonal)-Ratoon- wheat
Greengram- Autumn Sugarcane-ratoon-wheat	Paddy-Sugarcane-Ratoon- Finger millet
Maize- Autumn Sugarcane-ratoon-wheat	Paddy-Sugarcane-Ratoon- Wheat
Kharif Crops-Potato-Spring Sugarcane-ratoon-Wheat	Paddy-Sugarcane-Ratoon- gingelly (Til)
Kharif Crops-Mustard-Spring Sugarcane-ratoon-Wheat	Paddy-Sugarcane-Ratoon- urd
Kharif Crops-Pea/Coriander-SpringSugarcane-ratoon-Wheat	Cotton-Sugarcane-Ratoon–wheat
Kharif Crops-Wheat-late Planted Sugarcane-ratoon-Wheat	Sugarcane-Ratoon-Kharif rice-Winter rice

17. Ratooning in Sugarcane Cultivation

It is a practice of growing a crop from the stubbles of previous crop. The advantages of ratooning are; it saves cost on preparatory tillage and planting material. Apart from this, ratoon crop matures earlier and gives more or less same yield as regular sugarcane planting. The other benefit of ratoon is, it gets benefit of residual manure and moisture.

Disadvantages of Ratooning

Only one Ratoon should be taken because incidence of pests and diseases increases. The deterioration of soil is also a major concern.

18. Ratoon Management

18.1 Management of the Field after Harvest of the Plant Crop

Complete the following operations within 10 days of harvest of plant crop to obtain better establishment and uniform sprouting of shoots.

1. Incorporate the trash using trash shredder.

2. Follow stubble shaving with mechanical stubble shaver (RMD).
3. The gap in the ratoon sugarcane crop should be filled within 30 days of stubble shaving. The sprouted cane stubbles taken from the same field is the best material for full establishment. The next best method is gap filling with seedlings raised in poly-bags.
4. Apply manure and fertilizer as per recommendation.

18.2. Management of the Crop

1. 25 per cent additional N application on 5-7 days after ratooning.
2. Spray Ferrous sulphate at 2.5 kg/ha on the 15th day. If chlorotic condition persists, repeat twice further at 15 days interval. Add urea 2.5 kg/ha in the last spray.
3. Hoeing and weeding on 20th day and 40th to 50th day.
4. First top dressing on 25th day, 2nd on 45th to 50th day.
5. Final manuring on 70th to 75th day.
6. Partial earthing up on 50th day. If hoeing has been done two or three times up to 90th day, partial earthing up is not necessary.
7. Final earthing up on 90th day.
8. Detrashing on 120th and 180th day.
9. Trash twist propping on 180th day.
10. Harvest after 11 months.

19. Constraints in Sugarcane Cultivation

19.1 Sub-tropical Region

Long winters, low Sun shine hours, high temperature disparity, lack of early maturing high sucrose varieties, poor ratoon management, late planting, water logging, imbalance use of fertilizer, incidence of disease and pests, inadequate availability of quality seed, *etc.*

19.2 Tropical Region

Lack of irrigation in many states, inadequate availability of quality seed, incidence of disease and pests, poor ratoon management, *etc.*

20. Post-harvest Deterioration Losses

1. The sugar accumulated in the stem of sugarcane represents a balance between synthesis of sugar and its utilization. A well ripened harvested crop may lose its sugar within a few days after harvest, which tends to increase further due to high ambient temperature, pre-harvest burning, harvest and transportation injuries and microbial infestation. Sucrose losses after the harvest of sugarcane and during the subsequent milling operation are one of the most serious problems in many sugar processing mills in India. Staling beyond 24 hours results in considerable loss in cane

weight due to moisture loss and reduction in juice sucrose content due to inversion.

2. Such juice also creates problems in processing. The reduction in cane weight range between 7.4 and 17.0 per cent and sugar recovery by about 2.0 per cent at different places in India due to staling of cane for 96 hours. Thus post-harvest cane deterioration affects both growers because of loss in weight and sugar industry due to reduced recovery.
3. The stale canes reduce not only the recoverable sugar but also create losses by reducing mill and boiling house capacities. It also increases loss of sugar in molasses. Soon after the harvest of sugarcane, endogenous invertase enzyme gets activated and acts as a cause of deterioration. The other type of deterioration which is known as bio deterioration caused by microorganisms mainly *Leuconostoc* sp. (*L. mesenteroides* and *L. dextranium*) also takes place. These organisms convert sucrose into polysaccharides, such as dextran. Besides, loss of sucrose, the presence of dextran even in very small amount creates problem of filtration, clarification, crystallization and alters the shape of sugar crystals thereby affecting the quality of sugar.

21. Reasons for Stale Cane Crush

In sugarcane growing countries any of the following constraints seem to operate at farmers and factory level:

1. Absence of a proper varietal balance and scientific harvesting schedule based on cane maturity.
2. Extension of milling period during summer months when ambient temperature is high (>40°C).
3. Inordinate delay in transport of harvested cane from farmer's field/cane centres to the mills and lack of and efficient communication net/work.
4. Shut down and other unforeseen circumstances, such as labour scarcity and power interruption.
5. Mechanical harvesting of sugarcane (burnt crop) without proper and timely supply arrangements.

22. Strategies for Increasing Production and Productivity of Sugarcane

Sugarcane is long duration crop maturing in 12-14 months. The planting method vary place to place, therefore, the states has to encourage optimum use of water and space. Due to long crop duration inter cropping is a major tool to enhance higher returns to the farmers. The main strategies for increasing production and productivity of sugarcane in the country are as under:

- ☆ Popularization of new varieties by providing support for breeder seed production.

- ☆ Thrust for transfer of technologies through demonstration and training of farmers and extension workers.
- ☆ Production of quality planting materials including tissue culture plantlets.
- ☆ Training to the field functionary and farmers for popularizing various technology including Ratoon Management
- ☆ Demonstration on intercropping
- ☆ Establishment of bio-agent and tissue culture labs.

Chapter 23

Site Specific Nutrient Management: Approach for Precise Plant Nutrient Management

Amit Mishra[1], *Ipsita Kar*[2], *Jagganath Pathak*[1] *and A.K. Chaubey*[1]

[1]*Banda University of Agriculture and Technology, Banda – 210 001, U.P.*

[2]*Orissa University of Agriculture and Technology, Bhubaneswar – 751 003, Odisha*

Nutrient management is utmost important for eco-friendly sustainable improvement of soil and crop productivity. However, most farmers are still applying indiscriminate and imbalance nutrient thereby mining other nutrients particularly secondary and micro-nutrients. The phosphorus and potassium reserve are depleting and will be legitimate to assume that if constant heavy use of these fertilizer will continue then it will not only affect the food security but will also result in declined soil fertility. The adverse effect of ignorance of potassium application on crop growth and is visible after several years (Buresh *et al.*, 2010). Further, adverse effect on environment has apparently been visible such as ground water pollution and increase in N_2O concentration in atmosphere. Several strategies have been prepared for better managem ent of the nutrients and among them site specific nutrient management (SSNM) has commenced in 1990, which can be described as feeding of crop as and when needed. SSNM approach ensures balance and judicious use of plant nutrients, better crop production and reduction of cost of cultivation for sustainable soil health management and reduction in environment pollution.

Why SSNM

Soil Moisture Availability

It is one of the major components which ensures higher production of crops because improvement in fertilizer use efficiency (FUE) is directly related to availability of moisture crop production is affected by droughts as well as flood. Hence, maintenance of soil moisture is very important for crop production.

Soil Erosion

The loss of upper sheet of fertile soil due to erosion is very common in India, which deteriorate the fertility of the soil. Among the erosion, losses due to water erosion is maximum, which occurs due to faulty irrigation practice such as field to field irrigation and irrigated on unlevelled field *etc*. resulting in loss of top fertile soil, which contains maximum nutrients and organic matter. It is one of the reasons responsible for increasing the fertilizer requirement of the farmers.

Resource-Poor Farmers

Mostly in rainfed region, farmers are resource-poor, and they cannot afford crop loss. Which hinders adoption of innovative technology by farmers. Therefore, user-friendly and cost-effective technologies required for the rapid adoption.

Environmental Pollution

The imbalanced use of fertilizer not only increases the cost of cultivation but also affects the environment. For example, nitrous oxide is a greenhouse gas which is 310 times more potent than CO_2 and other hand its leaching pollute the groundwater. That affects the human and animal health.

Organic Matter

Organic matter is known as store house of the nutrients and plays vital role in soil fertility. However the declining trend of organic matter is observed in most of the Indian soils. Red and black soils and arid region soils are very poor in organic carbon content. Thus, use of organic manures is required for sustainable soil fertility.

Application of Nitrogenous Fertilizers

The nitrogen fertilizers are relatively cheap as compared to the other fertilizers. Moreover, their effect on crop is very prompt and pronounced than other macro, secondary and micro-nutrients. Therefore, farmers prefer to use of nitrogen fertilizers. The Ideal ratio of NPK application recommended is 4:2:1 but the currently NPK ratio is 6.8:2.7:1. This imbalance application of nitrogenous fertilizer is major challenge. On other side, in most of the rainfed region, suboptimal dose of fertilizer application is a major concern. Failure to apply K or P, on the other hand, can result in nutrient depletion and eventual loss in yield, which could take a number of seasons for detection (Abdulrachman *et al.*, 2006).

Way to Apply Precise Nutrients

It is possible to apply the nutrient precisely to the crops and improve fertilizer use efficiency by adopting various approaches of management approaches. The 4 R concept is globally accepted by the scientific community for judicious of fertilizers. It includes the following points:

Right Source

Ensure a balanced supply of essential nutrients, considering both naturally available sources and the characteristics of specific products in plant-available forms. Specifically, consider nutrient supply in plant-available forms to ensure the soil properties, and recognize the synergisms among elements.

Right Rate

Assess and make decisions based on soil nutrient supply and plant demand. Specifically, assess soil nutrient supply (including from organic sources and existing soil levels), assess plant demand, and predict fertilizer use efficiency.

Right Time

Assess and make decisions based on the dynamics of crop uptake, soil supply, nutrient loss risks and field operation logistics. Appropriately consider the timing of crop uptake, measure the dynamics of the soil's nutrient supply and recognize weather factors.

Right Place

Address root-soil dynamics and nutrient movement and manage spatial variability within the field to meet site-specific crop needs and limit potential losses from the field. Specifically, recognize root/soil dynamics, manage spatial variability issues, consider the tillage system, and limit potential off-field transport.

Right Tool

In addition to established 4 R now 5th R is needed to be added in this list, it suggests how much amount of fertilizer is needed to specify time and cultivar by the use of information communication technology (ICT). For ensuring implement of above all 4 R concepts, we need right tools. In recent years the information communication technologies have been improved. The use of these ICT tools is increasing in every sector including agriculture. These ICTs could be adopted as a tool for precise decision making and implementation.

Approaches for Precise Nutrient Management (Figure 23.1)

Soil Test-Based Crop Response (STCR)

Considering the importance of balanced nutrition and soil health, the ICAR initiated the long-term project on STCR. Under this project for the precise application of fertilizer, empirical equations for most of field and vegetable crops have been formulated. The essential basic data required for formulating fertilizer recommendation for targeted yield are (i) nutrient requirement in kg/q of produce,

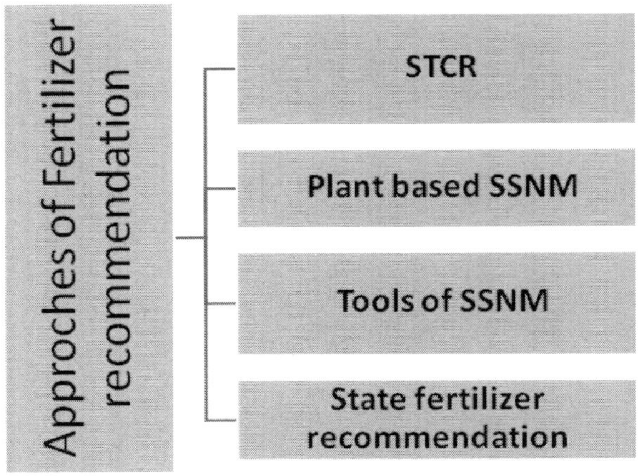

Figure 23.1: Various Approaches of Precisely Nutrient Recommendation.

grain or other economic produce (ii) the percent contribution from the soil available nutrients (iii) the percent contribution from the applied fertilizer nutrients.

The above mentioned three parameters are calculated as follows:

Nutrient requirement of N,P and K for grain production:

kg of nutrient per g grain = Total uptake of nutrient (kg) ÷ Grain yield (q)

$$\text{Contribution of nutrient from soil (CS\%)} = \frac{\text{Total uptake in control plot}\left(\frac{\text{kg}}{\text{ha}}\right)}{\text{Soil test values of nutrient in control plot (kg per ha)}} \times 100$$

Contribution of nutrient from fertilizer (CF per cent) = Total uptake of nutrients in treated plots- (Soil test value of nutrients in fertilizer treated plots x CS per cent)

$$\text{Contribution from fertilizer (CF\%)} = \frac{\text{CF}}{\text{Fertilizer dose (kg per ha)}} \times 100$$

Calculation of fertilizer dose: Above basic data are transformed into workable in prescription equation as follows:

$$\text{Fertilizer dose} = \frac{\text{Nutrient requirement in kg per ha of grain}}{\%\,\text{CF}} \times 100 \times \text{T} - \frac{\%\text{CS}}{\%\text{CF}} \times \text{Soil test value}$$

Source: Progress Report 2013-2016 AICRP on Soil Test Crop Response

Site Specific Nutrient Management

1. Plant based SSNM Approach

SSNM provides an approach for "feeding" crops with nutrients as and when needed. SSNM provides an approach for 'feeding' crop with nutrients as needed make optimal use of existing nutrients, such as from soil, residues, and manures (Buresh etal., 2015, Pampolino *et al.*, 2007). Apply fertilizer at the time and amount required by the crop. Apply P and K fertilizers based on crop need. The concept has been developed by systematic scientific studies by conducting several multi locations on station and on-farm trials through different research institutes. The trials mostly based on omission plot techniques. In this technique, the adequate amount of nutrients is applied except the nutrient of interest. The yield of omission plant indicates the nutrient supply potential of that soil. These trials also gathered a lot of information on local condition. The information includes cropping history and upcoming crop, variety, duration of variety, mode of sowing and other management practices, which influenced the fertilizer use efficiency. Moreover, the fertilizer recommendation is based on the specific location with respect to soil basic information related to fertilizer, relief and other constraint and also considered the cultivation practice of crops. On the basis of this information, the yield goal is decided and formulated the recommendation for fertilizer application for the particular crop of the specific location.

2. Tools of SSNM

In order to achieve efficient nitrogen management small handy tools have been developed and validated in different locations throughout the world. These instruments are quick and non-destructive. Several scientific study proved the effectiveness of these tools. These are based on optical sensors where as leaf colour chart depends upon colour of leaf. These instruments provide opportunity to manage the nitrogen based current crop performance. As the data is taken from the standing crops. Therefore, they are able to provide most correct reading.

i. Leaf Colour Chart (LCC)

LCC is very useful cheap handy tool for assessing nitrogen requirement in standing crop. It is made up of plastic plate having four strip of leaf colour ranging from yellow to dark green colour. LCC is used for rice, wheat and maize crop. Several studies on rice crop reported that at least 20 to 30 per cent saving on nitrogen fertilizer use and improvement in crop yield. The cost-effective ness is major advantage with LCC. Singh *et al.* (2012), studied the LCC approaches in several field trials in wheat crop and found significant improvement in grain yield and nutrient use efficiency by the LCC guided application of the nitrogen. Perceiving the importance of LCC several scientist recalibrated the requirement of nitrogen as per different varieties and locations.

Source: vcanr.edu/blogs/blogcore/postdetail(fm)potassium=14351.

ii. Green Seeker

Source: blogs.cornell.edu/agsci-interns/2014/08/06/thegreenseekerproject.

It is an optical sensor-based device which measures the biomass of the plant. Green plants absorb most of the red light and reflect most of the infrared light. The relative strength of the detected light is a direct indicator of the density of the foliage in the sensor's view. The denser and more vigorous the plant, the greater the difference is between the reflected light signals. The sensor displays the measured value on its LCD. NDVI can range from 0.00 to 0.99. The nitrogen fertilizer guided by green seeker has potential to increase the nitrogen use efficiency by 10 to 15 per cent by reducing the N application dose **(Ali MH *et al.,* 2015)**.

iii. Chlorophyll Meter (SPAD)

Source: blogs.cornell.edu/agsci-interns/2014/08/06/thegreenseekerproject.

SPAD (Soil-plant analysis development) is sensor-based device, which can be used directly on the field. The meter calculates a numerical SPAD value which is proportional to the amount of chlorophyll present in leaves.

These are handy devices, which have proven potential of site-specific nitrogen management. However, it is very difficult to convert these recommendations to large scale because farmers may not be able to purchase these costly devices for the cultivation. It is the major drawback of these types of the devices. The SPAD reading has been calibrated by several scientists for different locations and rice cultivars **Peng *et al.* (1996).** The SPAD reading 35 was found suitable for rice of South India **(IRRI-CREMNET, 1998, 2000, 2001).** However, slight variation in SPAD was also noticed by the scientist in respect to location, cultivars and other management practices. Hence, one standard reading may not be used for nitrogen recommendation irrespective of area and other management practice. Therefore, it is recommended to use SPAD meter for application of N dose of rice crop. Chlorophyll meter is also optimized by the research workers for other cereals like wheat and maize.

Rice Crop Manager

Computer or mobile phone-based tools are increasingly used to facilitate improved nutrient management practices in farmers' fields, especially in geographies where blanket fertilizer recommendations prevail. Crop Manager is developed by the International Rice Research Institute, Philippines (Bharti *et al.,* 2017). It is a computer-and mobile phone-based application that provides small-scale rice, rice-wheat, and maize farmers with the site- and season-specific recommendations for fertilizer application. The tool allows farmers to adjust the nutrient application to crop needs based on soil characteristics, water management, and crop variety on their farm. Recommendations are based on user-input information about farm location and management, which can be collected by extension workers, crop advisors, and service providers. The RCM has also been developed for stress-prone environments such as submergence, drought and saline-affected soils.

RCM has been very popular in Philippines, Vietnam and other south Asian countries. India, it is developed for eastern states like Bihar, Odisha and Eastern U.P in collaboration with ICAR- Institutes and State Agriculture Universities located in these regions. In Odisha, RCM has been adopted by Department of Agriculture and till now lakhs of recommendation generated from this tool. The software is freely downloadable at http://cropmanager.irri.org/home

Source: webapps.irri.o·g./n/

Nutrient Expert

Nutrient Expert® developed by International Plant Nutrition Institute based on principle of SSNM in cereal production systems. Nutrient Expert® is an interactive, computer-based decision-support tool that enables smallholder farmers to rapidly implement SSNM in their individual fields with or without soil test data. The algorithm for calculating fertilizer requirements was developed from on-farm research data and validated over 5 years of testing. The software is currently available without charge for wheat & maize systems in South Asia (http://software.ipni.net/article/nutrient-expert).

The mobile diffusion is very rapid in India, and these types of tools can be easily handled by extension workers, new age farmers and progressive farmers. Therefore, these type of technologies is freely available to the farmers. In future may be registered customer will get free information related to fertilizer on their mobile.

Conclusion

In an era of digitization, the mobile diffusion is very rapid in India in both urban as well as rural areas. Therefore, the web-based tool provides the opportunity to disseminate information to the farmers very quickly and precisely. Through mobile phone, anyone gets the feedback from the farmers very easily. As the digital Indian mission is going on most of the agriculture knowledge is now disseminated through using these platforms to the farmers. Even in the absence of these tools any extension worker can recommend the nutrient precisely, if he is having basic information on crop cultivation and soil fertility map. It is an urgent need to replace the blanket recommendation of fertilizer by SSNM. The nutrient budgeting for the whole crop-livestock system is key to the success of SSNM. However, validation of tools/ technologies in a particular location is utmost necessary before initiating mass scale recommendation.

References

Abdulrachman S, Witt C, Buresh RJ (2006) The need for potassium fertilization in rice and experiences from a long term experiment in Indonesia. e-ifc no. 10, International Potash Institute. www.ipipotash.org/eifc/2006/10/6. Accessed 25 January 2019

Ali A.M., Thind H.S., Singh Varinderpal and Singh Bijay (2015) A framework for refining nitrogen management in dry direct-seeded rice using GreenSeeker optical sensor. Computers and Electronics in Agriculture, Vol: 110: 114-120

Bharti Preeti, Annand Priyanka, Mishra Amit and Sharma Sheetal (2017) An app tailor-made for India's rice farmers Rice today. Vol. 16:4: 20-22.

Buresh, R.J., Pampolino, M.F. and Witt, C. 2010. Field-specific potassium and phosphorus balances and fertiliser requirements for irrigated rice based cropping systems. *Plant and Soil* **335,** 35-64.

Buresh, R.J., Witt, C., Ramanathan, S., Mishra, B., Chandrasekaran, B. and Rajendran, R. 2005. Sites specific nutrient management: Managing N, P, and K for rice. *Fertiliser News* **50(????),** 25-28.

IRRI-CREMNET. 1998. *Progress Report – 1997*. International Rice Research Institute - Crop and Resource Management Network, Los Baños, Philippines.

IRRI-CREMNET. 2000. *Progress Report - 1998-1999*. International Rice Research Institute -Crop and Resource Management Network, Los Baños, Philippines.

IRRI-CREMNET. 2001. *Progress Report - 1999-2000*. International Rice Research Institute -Crop and Resource Management Network, Los Baños, Philippines.

Pampolino M.F., Manguiata I.J., Ramanathan S., Gines H.C., Tan P.S.,Chi T.T.N.,Rajendran R., and

Buresh R.J. (2007) Environmental impact and economic benefits of site-specific nutrient management (SSNM) in irrigated rice systems. *Agricultural Systems*, 93: 1-3:1-24.

Peng, S., Garcia, F.V., Laza, R.C., Sanico, A.L., Visperas, R.M. and Cassman, K.G. 1996. Increased nitrogen use efficiency using a chlorophyll meter in high-yielding irrigated rice. *Field Crops Research* 47, 243-252.

Progress Report 2016, All India Coordinated Research Project on Soil Test Crop Response. ICAR-Indian Institute of of Soil Science, Bhopal, India

Singh Varinderpal, Singh Bijay, Singh Yadvinder, Tind H.S., Singh G. Kaur S., Kumar Ajay, and Vashistha Monika (2012). Establishment of threshold leaf colour greenness for need-based fertilizer nitrogen management in irrigated wheat (*Triticum aestivum* L.) using leaf colour chart. *Field Crop Research*, 130: 109-119.

Chapter 24

Protected Cultivation of High Value Horticultural Crops

Ajay Kumar Singh[1], Rakesh Kumar[2] and K.S. Tomar[2]

[1]Associate Professor, [2]Assistant Professor
(Floriculture and Landscape Architecture)
College of Horticulture, BUAT, Banda – 210 001, U.P.

Introduction

India has varied climatic conditions in different regions, so the greenhouse and the supporting facilities have to be developed accordingly. The local climatic conditions play a major role in design of the greenhouse structures and cladding materials. The cost does not indicate specific make of controlled environment, since it depends upon climatic parameters such as sun light, temperature, relative humidity, *etc.* Protected cultivation includes hardening structures, tunnels, shade nets, ventilated structures, polyhouses, *etc.* They are being used for the following on limited scale for raising early season vegetables and flower nursery, production of early vegetable crops, extension of growing season, vegetable production during frozen winters in high altitude area, protection of valuable germplasm and cultivation of cucurbits, brinjal, capsicum and certain flowering annuals. There is a gross increase in yield and quality of produce under protected cultivation. There are different types of greenhouses such as naturally ventilated greenhouse, greenhouse with fan and pad cooling system, hi-tech greenhouse and bamboo/ wooden structure greenhouse for growing vegetables (tomato, capsicum, coloured capsicum, parthenocarpic cucumbers, cauliflower, cabbage, okra, bitter gourd, onion, coriander, spinach *etc.*), fruits (strawberry, papaya *etc.*) and flowers (rose, chrysanthemum, carnation, gerbera, lilium, orchid *etc.*).

Historical Background

Since time immemorial, man has learnt how to grow plants under natural environmental conditions. In some of the temperate regions where the climatic conditions are extremely adverse and no crops can be grown, man has developed methods of growing some high value crops continuously by providing protection from the excessive cold, which is called as Greenhouse Technology. So, Greenhouse Technology is the technique of providing favorable environment condition to the plants. It is rather used to protect the plants from the adverse climatic conditions such as wind, cold, precipitation, excessive radiation, extreme temperature, insects and diseases. It is also of vital importance to create an ideal micro climate around the plants. The greenhouse technology is more than 200 years old and Europeans were the pioneers in this field. Later, with the advent of plastics during World War II; a new phase of greenhouse technology emerged. There are more than 50 countries now in the world where cultivation of crops is undertaken on a commercial scale under cover. The Netherland alone has around 9000 greenhouse enterprises that operate over 10,000 ha of area and employ about 150,000 workers efficiently. The area under greenhouses in Spain has been estimated to be around 25,000 ha and Italy 18,500 ha used mostly for growing vegetable crops like watermelon, capsicum, strawberries, beans, cucumbers and tomatoes. In Spain simple tunnel type greenhouses are generally used without any elaborate environmental control equipments mostly using UV stabilised polyethylene film as cladding material.

Japan has more than 40,000 ha under greenhouse cultivation of which nearly 7500 ha is devoted to only fruit orchards. Greenhouses in Japan are used to grow wide range of vegetables and flowers with a considerable share of vegetable demand being met from greenhouse production. Even a country like South Korea has more than 21,000 ha under greenhouses for production of flowers and fruits. Thus, greenhouses permit crop production in areas where winters are severe and extremely cold as in Canada and USSR, and also permit production even in areas where summers are extremely intolerable as in Israel, UAE and Kuwait. Greenhouses in Philippines make it possible to grow crops in spite of excessive rains and also in moderate climates of several other countries. Thus, in essence greenhouse cultivation is being practiced and possible in all types of climatic conditions.

Status and Scope in India

In India use of greenhouse technology started only during 1980's. In India, the technology is still in its nascent stage. The area under greenhouse cultivation in India, as reported by National Horticulture Mission, Government of India was about 2000 ha including 500 ha net house, shade nets 1500ha greenhouse in the year 2005 which has grown up to 80,795.81 ha by the end of 2012.

Greenhouses are being built in the Ladakh region for extending the growing season of vegetables from 3 to 8 months. In the North-East, greenhouses are being constructed essentially as rain shelters to permit off-season vegetable production. In the Northern plains, seedlings of vegetables and flowers are being raised in the greenhouses either for capturing the early markets or to improve the quality of the seedlings. Propagation of difficult-to-root tree species has also been found to be very encouraging. Several commercial floriculture ventures are coming up in Maharashtra, Tamil Nadu and Karnataka states to meet the demands of both domestic and export markets. The commercial utilization of greenhouses started from 1988 onwards and now with the introduction of Government's liberalization policies and developmental initiatives, several corporate houses have entered to set up 100 per cent export oriented units. In just four years, since implementation of the new policies in 1991, 103 projects with foreign investment of more than Rs.80 crores have been approved to be set up in the country at an estimated cost of more than Rs.1000 crores around Pune, Bangalore, Hyderabad and Delhi. Thus the area under climatically controlled greenhouses of these projects is estimated to be around 300 ha. Out of which many have already commenced exports and have received very encouraging results in terms of the acceptance of the quality in major markets abroad and the price obtained.

Basic Principle of Protected Cultivation

A greenhouse is generally covered with a transparent material such as polythene or glass. Depending upon the cladding material and its transparency major fraction of sunlight is absorbed by vegetable crops and other objects. These objects in greenhouse in turn emit long wave thermal radiations for which cladding material has lower transparency. With the result, solar energy is trapped and raises the temperature inside the greenhouse. This is popularly known as greenhouse effect. This rise in temperature in greenhouse is responsible for growing vegetable in cold climates. During summer months, air temperature in greenhouse is to be brought down by providing cooling device. In commercial greenhouses besides temperature-controlled humidity, carbon dioxide, photoperiod, soil temperature, plant nutrients *etc.* facilitate round the year production of desired vegetable crops. Controlled climatic and soil conditions provide an opportunity to the vegetable crops to express their full yield potentials.

Advantages of Protected Cultivation

i) The yield may be 10-12 times higher than that of outdoor cultivation depending upon the type of greenhouse, type of crop and environmental control facilities.

ii) Reliability of crop increases under greenhouse cultivation.

iii) Ideally suited for vegetables and flower crops.

iv) Year round production of floricultural crops.

v) Off-season production of vegetable and fruit crops.

vi) Disease-free and genetically superior transplants can be produced continuously.

vii) Efficient utilization of chemicals, pesticides to control pest and diseases.

viii) Water requirement of crops very limited and easy to control.

ix) Maintenance of stock plants, cultivating grafted plant-lets and micro propagated plant-lets.

x) Hardening of tissue cultured plants

xi) Production of export quality produce free of blemishes.

xii) Most useful in monitoring and controlling the instability of various ecological system.

xiii) Modern techniques of Hydroponic (Soil less culture), Aeroponics and Nutrient film techniques are possible only under greenhouse cultivation.

Classification of Greenhouse/Polyhouse

Greenhouse structure of various types is used for crop production. Although, there are advantages in each type for a particular application, in general there is no single type greenhouse, which can be constituted as the best. Different types of greenhouses are designed to meet the specific needs. The different types of greenhouses based on shape, utility, material and construction are briefly given below:

1. Greenhouse Type Based on Shape

For the purpose of classification, the uniqueness of cross section of the greenhouses can be considered as a factor. The commonly followed types of greenhouses based on shape are:

a) Lean to type greenhouse.

b) Even span type greenhouse.

c) Uneven span type greenhouse.

d) Ridge and furrow type.

e) Saw tooth type.

f) Quonset greenhouse.

g) Interlocking ridges and furrow type Quonset greenhouse.

h) Ground to ground greenhouse.

2. Greenhouse Type Based on Utility

Classification can be made depending on the functions or utilities. Of the different utilities, artificial cooling and heating are more expensive and elaborate. Hence based on this, they are classified into two types:

a) Greenhouses for active heating.

b) Greenhouses for active cooling.

3. Greenhouse Type Based on Construction

The type of construction predominantly is influenced by structural material, though the covering material also influences the type. Higher the span, stronger should be the material and more structural members are used to make sturdy tissues. For smaller spans, simple designs like hoops can be followed. So based on construction, greenhouses can be classified as:

a) Wooden framed structure.

b) Pipe framed structure.

c) Truss framed structure.

4. Greenhouse Type Based on Covering Material

Covering materials are the important component of the greenhouse structure. They have direct influence on greenhouse effect, inside the structure and they alter the air temperature inside. The types of frames and method of fixing also varies with covering material. Hence based on the type of covering material they may be classified as:

a) Glass glazing.

b) Fibre glass reinforced plastic (FRP) glazing

   i. Plain sheet

   ii. Corrugated sheet.

c) Plastic film

   i. UV stabilized LDPE film.

   ii. Silpaulin type sheet.

   iii. Net house.

d) Based on the cost of construction involved (which includes various factors mentioned from a to c)

   i. High cost Green House

   ii. Medium cost Green House

   iii. Low cost Green House

Mode of Protected Cultivation

1. Low-Cost Greenhouse/Polyhouse

The low cost polyhouse is a zero-energy chamber made of polythene sheet of 700 gauge supported on bamboos with sutli (ropes) and nails. It will be used for protecting the crop from high rainfall. Its size depends upon the purpose and availability of space. The structure depends on the sun for energy. The temperature within polyhouse increases by 6-10°C as compared to outside temperature. In UV stabilized plastic film covered pipe framed polyhouse, the day temperature is higher and night temperature is lower than the outside. The solar radiation entering the polyhouse is 30-40 per cent lower than that reaching the soil surface outside.

2. Medium-Cost Greenhouse/Polyhouse

With a slightly higher cost, a Quonset-shaped polyhouse (greenhouse) can be framed with GI pipe (class B) of 15 mm bore. This polyhouse will have a single layer covering of UV-stabilized polythene of 800 gauge. The exhaust fans are used for ventilation. These are thermostatically controlled. Cooling pad is used for humidifying the air entering the polyhouse. The polyhouse frame and glazing material have a life span of about 20 years and 2 years, respectively.

3. High Cost Greenhouse/Polyhouse

It is constructed on the structure (frame) made of iron/aluminum structure, designed domed shaped or cone shaped (as per choice). Temperature, humidity and light are automatically controlled as per requirement of the users. Floor and a part of walls are made of concrete. It is highly durable, about 5-6 times costlier, required qualified operator, proper maintenance, care and precautions while operating. The low and medium-cost greenhouses have wide scope in production of domestic as well as export-oriented vegetables. NEH region recorded highest rainfall in the world. The duration of rainy season is also wide (April-October). During this period, growing of vegetables such as cabbage, cauliflower, broccoli, tomato, brinjal and French bean in open conditions is very difficult. Severe attacks of pest and diseases occur due to heavy rains. So, growing of vegetable crops in low cost polyhouse during this period is very profitable. Control of disease and pest in polyhouse is also easy.

Other Plant Protection Structures

1. **Plastic low tunnels:** Plastic low tunnels are miniature form of greenhouses to protect the plants from rains, winds, low temperature, frost and other vagaries of weather. The low tunnels are very simple structures requiring very limited skills to maintain, easy to construct and offer multiple advantages. For construction of low tunnels, film of 100 micron would be sufficient. The cost of a 100-micron thick film would be about Rs.10/m^2.
2. **Hot bed and Cold frame:** Hot bed is made of wood, cement or brick stone and is prepared above ground with artificial heating. Used for growing vegetables in winter and provide an early start of crops like – tomato, cucumber, okra, pepper, cabbage, musk melon and water melon *etc.* Cold frames are similar to hot beds except for absence of any form of artificial heat and having plastic on top for tapping solar energy or natural heat.
3. **Net houses:** Net houses are used for raising vegetable crops in high rainfall regions. Roof of the structure is covered with suitable cladding material. Sides are made of wire mesh of different gauges. Such structures are useful for NEH region.
4. **Cloches:** It is a protective enclosure, consisting of a structural frame and a transparent glazing material for one plant. It is used to provide protection to potted plants and young transplants in fruit orchard and forests.
5. **Rain shelter:** Most suited protective structure in high rainfall states like – Assam and Kerala. Mostly even span low cost green house structures are used as rain shelter.
6. **Floating row cover:** It is a plastic film fabric used without any mechanical support in the form of hoops to protect crops from insect vectors, practiced for winter production of tender warm season crops to enhance early maturity, improve quality and extent the growing season.

7. **Trench cultivation:** Trench cultivation, otherwise called as "under ground greenhouse technology" is a simple and economically viable technology for growing vegetable during winter. This technology is best utilized by farmers of Ladakh and Leh region and Jammu and Kashmir for extending vegetable production from three to eight months. Important vegetables which can be grown in trenches are beet leaf, spinach, Chinese cabbage, celery, parsley, coriander *etc.*

Mulching

Mulching is the practice of covering soil around the plants with organic or synthetic materials to make conditions more favorable for plant growth, development and crop production.

Types of Mulches

1. **Loose organic and non organic mulches:** Cocoa bean shell, Coffee ground and tea leaves, Dust mulches, Feather mulches, Flax, Glass clippings, Gravels, Hop waste, Leaves and leaf mould, Manures (for vegetables like – asparagus, leek and potato), Mushroom compost, Peat, Sand, Sawdust, Stones, Straw *etc.*
2. **Inorganic or Synthetic mulches**- Plastic mulches are more expensive, non-degradable (commonly used) which are of following types:
   I. **Black plastic mulch-** It is most popular, retard weed growth, warm soil during winter and encourage plant growth.
   II. **Clear or transparent plastic mulch-** It warms soil environment very fast often referred as mini green house effect; drastically temperature rise may adversely affect plant growth. It is mostly used in soil solarization
   III. **White plastic mulch-** It reflects more light, keep soil temperature down used during hot summer condition. It also repels some insects
   IV. **Degradable mulch-** It degrade after having received a critical amount of sun light, film eventually break down into small flakes and disappear in the soil. *e.g.* Biolan, Plastigone
   V. **Selective permeable/reflective (colored) plastic mulch-** It allows certain wavelength of light to pass through mulch and warm the soil beside discouraging weeds growth. Aluminum faced plastic interfere aphid movement, yellow plastic mulch repel white fly. It give marketable produce over longer periods of time.
   VI. **Photo selective plastic mulch-** It recently developed also known as infra-red transmitting (IRT) mulch. Transmit higher proportion of solar infra-red radiation and blocking most photo-synthetically active radiation so retard growth of weeds. Crops RTH 7-10 days earlier than using black polythene as mulch.

Advantages of Plastic Mulch

1. Increase soil temperature (black plastic- 4-5 °F, clear film – 8-10 °F) upto 2 inch depth of soil
2. Increase plant growth- higher CO_2 conc. due to chimney effect
3. Earlier harvesting in black plastic (7-10 days) and clear mulch (14-21 days advances)
4. Reduce evaporation, weed problems and fertilizer leaching
5. Eliminating root pruning
6. Facilitate cleaner vegetable cultivation
7. Mulches aid fumigation- necessary to control soil born diseases (*Fusarium* wilt)
8. Helps in insect management

Disadvantages

1. Encourages some pernicious forms (like sludge and snails), fungal and other diseases.
2. Impermeable mulch may restrict soil aeration.
3. Removal of non degradable plastic mulches is essential

Micro-irrigation, Fertigation and Nutrient Need Diagnosis

Micro-irrigation is an advanced methods of application of water, by which only required quantity of water can be supplied to the root zone of soil. Israeli engineer, Symcha Blass developed this technique first. Now it becomes one of the most important tools for protected vegetable cultivation. This technique has following advantages

Advantages

1. Save water by 50-70 per cent
2. Reduce labour and energy costs
3. Weed infestation negligible (less wetting soil)
4. Reduce adverse effect of saline soil
5. Save fertilizers by 40-60 per cent (checking leaching loss)
6. Eliminate soil erosion
7. Improve quality and quantity of produce/crop
8. Reduce incidence of insect pest and diseases

Constraints

1. High initial investment
2. Frequent clogging of drippers (algae, salt accumulation, foreign particles)
3. Require technical expertise

4. Environment pollution due to non-degradable plastic
5. Not suitable for frost protection or for cooling during hot weather period or for supplemental irrigation of large areas.

Components of Drip Irrigation

1. Water emitting device- drippers supply required amount of water
2. Distribution lines- made of PVC and HDPE
3. Fittings- used for connecting PVC and HPDE
4. Control head system- consists of pump, primary filter, pressure regulator, water meter and fertilizer injectors
5. Filter system- screen filter frequently used
6. Other accessories- starter, plug, joints, tees *etc.*

Factors Limiting Large Scale Adaptation of Drip Irrigation System

1. Low level of awareness, inadequate knowledge about its application, utility and maintenance
2. High initial cost and non-availability of basic finance, govt. incentives
3. Operational problems like:
   i) Non-availability of spares/micro tubes
   ii) Leakage due to damage of pipes or drippers
   iii) Clogging of drippers

Fertigation

It implies the application of fertilizers through pressurized systems; thus forming nutrient containing irrigation water desired concentration as per the crop need. Standard practice of fertilizer application in greenhouse.

Advantages

1. Fertilizer completely dissolved in water so that there will be no clogging of drippers and emitting pipes
2. Improving growth and productivity due to correct fertigation schedule
3. Nutrients applied as and when required to meet specific requirement of crop
4. No loss of nutrients (applied at root at rhizospere) through volatilization
5. Water soluble nutrients can be supplied
6. Safer method of application of fertilizer; no leaf burning chance
7. Save water, labour, time, fertilizer (25-50 per cent), energy and cost involvement
8. Incorporation of fungicides, pesticides and nematicides also possible

Constraints

1. Non availability of water soluble fertilizer at reasonable rates
2. Lack of R&D efforts for developing fertilizers for fertigation
3. Inadequate policy of Govt. for promoting this technology

Forms of Fertilizer Application

A) **Liquid fertilizer**- These are the solutions which contain one or more plant nutrients in liquid or suspension form.

B) **Water soluble fertilizer**- These are the fungicides which are completely soluble in water containing two or more major as well as minor nutrients.

C) **Suitable conditions for fertigation**- It is most suitable method of drip irrigation system with rain water. Water having pH 5.5 to 6.5, EC less than 0.1, low level of carbonate, bicarbonate, sodium, chlorine is ideal.

Utilization of Protected Cultivation

1. Raising Off Season Nurseries

The cost of hybrid seeds is very high. So, it is necessary that every seed must be germinated. For 100 per cent germination, it requires controlled conditions. The cucurbits are warm season crops. They are sown in last week of March to April when night temperature is around 18-20°C. But in polyhouse,their seedlings can be raised during December to January in polythene bags. Planting of these seedlings during end of February and first week of March in the field, the crop could be taken in one and one and a half months in advance than the normal method of direct sowing. This technology fetches bonus price due to marketing of produce in the off-season. Similarly, the seedlings of tomato, chilli, capsicum, brinjal, cucumber, cabbage, cauliflower and broccoli can be grown under plastic cover protecting them against frost, severe cold and heavy rains. The environmental condition, particularly increase in temperature inside polyhouse hastens the germination and early growth of warm season vegetable seedlings for raising early crops in spring summer. Vegetable nursery raising under protected conditions is becoming popular throughout the country especially in hilly regions. Management of vegetable nursery in protected structure is easier and early nursery can be raised. Needless to emphasize, this practice eliminates danger of destruction of nurseries by hail storms and heavy rains. Protection against biotic and abiotic stresses also becomes easier.

Requirement for Raising Vegetables Seedling by Hi-tech Nursery

1. Protected structure
2. High yielding hybrid varieties
3. Soilless media
4. Multicelled plastic plug-trays
5. Fertigation system

Media for Raising Seedlings

1. **Peat:** It is remains of aquatic marsh bog or swamp vegetation in partial decomposed state. It has high moisture holding capacity.
2. **Coco peat**: It is by product of coconut husk. Due to honey comb structure in fibres, it is best for providing aeration with improved drainage and porosity.
3. **Perlite**: It is grey-white silicacious material of volcanic origin. It is neutral in pH.
4. **Vermiculite:** This is a micaceous mineral. Chemically it is a hydrated magnesium –aluminum iron Silicate. It is light in weight and provides magnesium and potassium to enrich the media and has good water holding capacity.
5. **Rockwool:** It is produced by burning a mixture of rock, basalt and limestone.

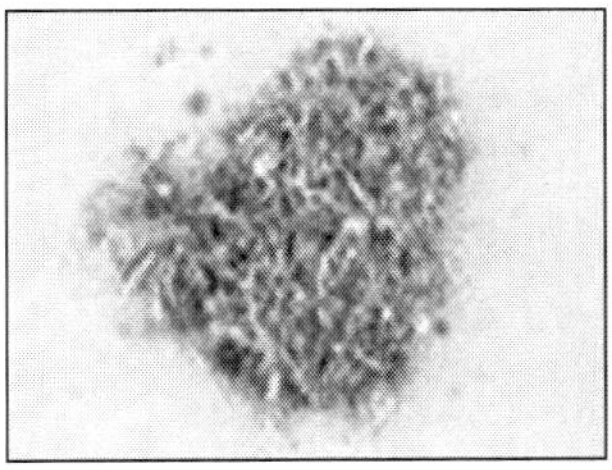

Peat

Coco Peat

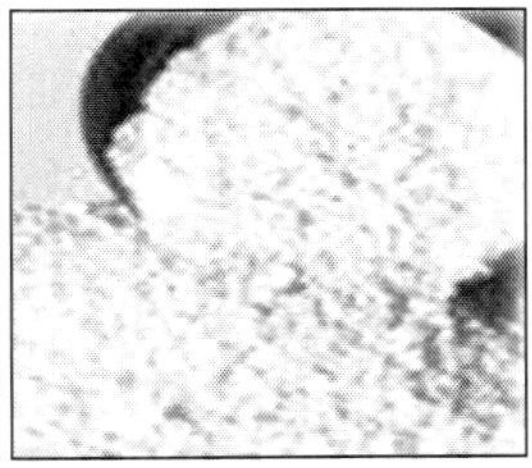

Perlite

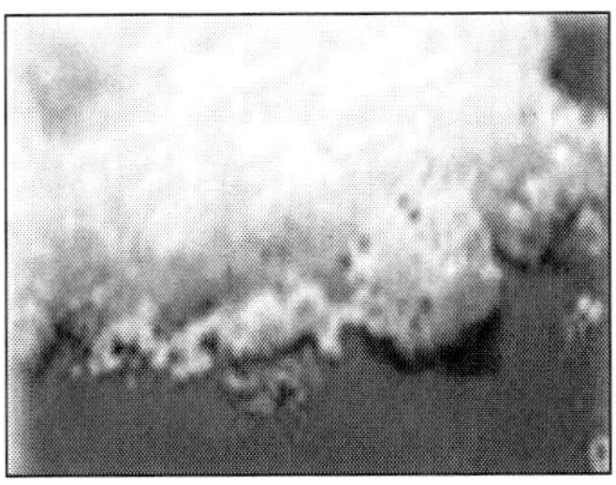

Hydrogel

Rockwool

Vermiculite

Advantages of Soil-less Media

1. **Uniformity of mix**-The physical and chemical properties of a mix are uniform through the mix, a condition not found in field soil. The homogeneity of the medium makes it possible for plants to grow and develop uniformity.
2. **Ease of handling**-Mixes are light in weight and easy to transport. Ingredients are easy to scoop and move around during manual preparation.

3. **Versatility-**Mixes can be made to order (custom mixed) for specific needs. They can be used to amend the soils in field by mixing nursery beds *etc.* They are convenient for use in container plants.
4. **Sterility-**Mixes are initially free from diseases and insect pest. the seed germination, therefore, is less prone to diseases such as damping off.
5. **Good drainage and moisture retention-**A mix can be custom made to provide an appropriate degree of drainage and moisture retention.
6. **Convenience of use-**Mixes is available in ready to use form.

Advantages of Hi-tech Nursery Transplants over Field Grown Transplants

1. They can be produced more uniformly than field grown plants.
2. Favorable growing condition can easily be provided for raising seedlings successfully.
3. Transplants often result in better stands.
4. Higher quality and uniformity of transplants.
5. Earlier establishment and earlier harvest after transplanting.
6. More suitable for mechanical transplanting.
7. Reliable planning of production.
8. Their growth can be controlled more easily through fertility and water management and they can be held longer and harvested when needed.
9. Stockier plants can be produced because container grown plants are less crowded and healthier.
10. Container grown transplants have a media-enclosed root ball, which retains moisture and root integrity at transplanting, reducing transplanting shock.
11. Easy and convenient caring of young tender seedlings growing in small areas against biotic and abiotic stresses.
12. Higher germination rate and more high quality transplants can be produced in greenhouse under controlled environment conditions.
13. Cost associated with raising transplants can be easily be recouped.
14. Less seed is used.
15. Higher survival rate of seedlings, because of reduced damage in transplanting and lower incidence of disease.

2. Vegetable Seed Production

Seed production in vegetables is the limiting factor for cultivation of vegetables in NEH region of India as well as in India. The vegetables require specific temperature and other climatic conditions for flowering and fruit setting. Seed production of brinjal, capsicum, cauliflower and broccoli is very difficult in open conditions in this area due to high rainfall at maturity stage. To reduce such micro climatic condition, a protected environment is essential. Therefore, the seed production of highly remunerative crops namely tomato, capsicum and cucumber is performed under protected environments. The maintenance and purity of different varieties/ lines can be achieved by growing them under greenhouse without giving isolation distance particularly in cross-pollinated vegetables namely onion, cauliflower and cabbage. Hence, vegetable production for domestic consumption and export in low and medium cost greenhouse is a technical reality in India. Such production system has not only extended the growing season of vegetables and their availability but also encouraged conservation of different rare vegetables.

3. Hybrid Seed Production

In 21st century, protected vegetable production is likely to be commercial practice not only because of its potential but out of sheer necessity. In vegetable production hybrids seeds, transgenic, stress resistant varieties, micropropagated transplants, synthetic seeds are likely to replace conventional varieties. Protected environments will be helpful in production of hybrid seeds of cucumber and summer squash by using gynoecious lines. Gibberlic acid is used to maintain such lines followed by selfing. The desired pollen can be used for production of hybrid seed of cucumber. Similarly in summer squash use of Ethophan in inducing female flower at every node would help in the hybrid seed production by using desired pollen parent.

4. Maintenance and Multiplication of Self Incompatible Line for Hybrid Seed Production

In case of cauliflower, there is problem of maintaining and multiplication of potential self-incompatible lines for the production of F1 hybrid seed. Temporary elimination of the self-incompatibility with the use of CO_2 gas has solved this problem. For this purpose, the self-incompatible line is planted in a greenhouse and bees are allowed to pollinate the crop when it bloom. Then keeping the greenhouse closed tightly, within 2-6 hours of pollination, it is treated with 2-5 per cent CO_2 gas which allows successful fertilization by temporarily eliminating the self-incompatibility

5. Polyhouse for Plant Propagation

Asparagus, sweet potato, pointed gourd and ivy gourd are sensitive to low temperature. The propagating materials of these vegetables can be well- maintained under polyhouse in winter season before planting their cuttings in early spring-summer season for higher profit.

6. Off-Season Vegetable Cultivation

In Ladakh, till early 1990s cucurbits cultivation in open was impossible. But growing seedlings in poly pouch under polyhouse during April-May and transplanting them in open field ith plastic mulch made it possible to grow almost all cucurbits in Leh. This has not only improved vegetable basket in the area but also added variety to food basket of local inhabitants and soldiers. Commercial production of cucurbits in cold desert of India is now possible through protected cultivation. Sarda melon imported in large quantity in the country can be produced in these areas with ease. Production of off-season (August and September) muskmelon and watermelon in open fields has also become possible. An early crop of cucurbits like squash and longmelon is also grown in polyhouses.

Production Practices of High Value Vegetable Crops

Bell Peppers

Peppers require comparatively long growing period to reach transplant size and normally, transplants of 35 to 45 days old are used. Paired row method of planting is adopted with 60 cm x 30 cm spacing and between paired rows, a spacing of 90 cm is maintained, accomodating about 44000 plants per ha. Compared to other vegetables, it grows well with higher humidity conditions in the range of 60 – 70 percent. Capsicum plants are pruned to retain four stems. Pruning is done at weekly intervals, starting from 15 to 20 days after transplanting (DAT). Capsicum plants show dichotomous branching. The tip of the plant splits into two at 5th or 6th node, 15 to 20 days after transplanting. These two branches again split into two giving rise to four branches by 25 to 30 DAT (Hebber *et al.*,, 2011). Only these four main branches are maintained till the end of the crop. At every node the tip split into two giving rise to one strong branch (the branch with thick stem and larger leaf) and one weak branch (the branch with thin stem and smaller leaf). The weak branch i8s removed, retaining the strong branch. This operation needs to be done once in a week. From 4th month onwards, the pruning operation will be done once in 10 days. All the shoots appearing at the base of the plants are removed as they are not productive. One to two fruit will set per node. Plants are often vibrated or trellis wires tapped to ensure proper pollination and fruit set.

Tomatoes

Tomatoes selected for greenhouse production are indeterminate types. Indeterminate tomatoes continue to grow and set fruit throughout their life cycle. Varieties selected for greenhouse production fall into two broad categories: large beefsteak or slicing types and cluster or hand types. The larger types are harvested singly. The cluster or hand types are harvested in clusters of four to seven fruits and are sold with the cluster stem still attached. Seedlings for transplanting are ready around 24 to 26 days after seedling. Make sure that the medium the plants will be grown in is thoroughly moistened with water or dilute nutrient solution. If growth bags of perlite are used for production, they should be moistened as described above, and the irrigation emitters should be placed in each bag before slits are cut to allow excess solution to drain away. Paired row method of planting is adopted with 60 cm x 45 cm spacing and between paired rows a spacing of 90 cm is maintained accommodating about 30,000 plants per hectare. Indeterminate tomato varieties are trained to a single stem and supported by twine. The plants must be pruned and the fruit thinned throughout the crop cycle. As the plants grow up the string, lateral branches are removed from leaf axils while the main stem is allowed to grow continuously. Fruits must be thinned as they forming the flower clusters. The variety will determine how many fruits should be left at each flower cluster. Normally, 4-5 fruits per cluster are retained. The flowers must be vibrated

or shaken each day that flowers are produced to ensure pollination. The best time to do this necessary operation is between 10 a.m. and 3 p.m. During overcast or rainy days, pollen will not be dry enough to adequately pollinate the crop; therefore, plants must be shaken during sunny dry days.

Cucumbers

Cucumber varieties grown in greenhouses are known as European or English or baby cucumbers. These special cucumbers are parthenocapic, that is, they produce fruit without the need for pollination. In fact, care should be taken to avoid pollination by non-parthenocarpic cucumbers because this will result in bitter and misshapen fruit. These varieties are gynoecious and multi pistillate. Cucumbers are a warm season crop; therefore, greenhouse temperatures should not be allowed to drop below 18°C. Seed germination can occur with media temperatures above 16°C, but optimal germination occurs at 30°C to 35°C. Typically, cucumbers will be germinated in flats of rock wool or foam blocks, usually with a single seed per well to ensure plants can be transplanted with minimal root damage. Plants should be transplanted to final spacing in the greenhouse when they are large enough to handle without damage. Thay will usually be ready for transplanting in two to three weeks under optimal conditions. Cucumbers are vining crops, so they are grown with a trellis system. As the plant is trained, all fruits below 24 to 30 inches are removed to maximize foliar growth. Above this height, fruits are allowed to set. Several different trellis systems have been developed to produce greenhouse cucumbers. These systems, using either one or two wires strung approximately 8 feet above the crop, are referred to as single and V-cordon systems. Both single and double rows of plants can be used with the single cordon system. In either case, the plants are trained up a string to a suspended wire where the plant is draped over the wire. At this time, the plant may be pinched and two laterals allowed growing down to the ground, or the primary stem may be allowed to continue growing to the ground. In either case, all other laterals are removed as they appear and at least one fruit is allowed to develop at each node. In the V-cordon system, two wires are suspended so that the plants can be alternately strung to one wire or the other. This results in a V-shaped system of plants growing to the respective wires. In all other ways, they are treated the same as the single wire system in terms of pruning laterals and draping over the wire. In addition, the in-row spacing will differ between the two systems. In the single cordon system, plants are spaced 12 to 18 inches apart. If a single wire with a double row is used, the plants are spaced 60 cm apart with 60 cm between the double rows.With the V-cordon system, the plants are spaced 30 cm apart with plants alternatively strung to one wire or the other. Between- row spacing for the double-row single-wire system is 120 cm to 150 cm; for single-row single-wire system, 90 cm to 120 cm. With the V-cordon system, between-row spacing is 120 cm. Fruit matures continuously as the plants grow. Flower opening to harvestable fruit will take between 10 and 14 days. Therefore, harvesting is a continual process, usually done three times a week. Harvested fruit must be protected from desiccation and bruising because the skin is rather tender and prone to water loss. Greenhouse cucumbers are individually shrink-wrapped to restrict desiccation and extend shelf life.

Economics of Greenhouse Vegetable Production

Greenhouses are expensive to build and operate. A commercial greenhouse of one acre with shade net and foggers, fertigation and ventilation system will cost between Rs. 30 to 35 lakhs to erect an equipement on the other hand, one acre sized low cost naturally ventilated simple greenhouses- like stone or wooden pole and plastic or nylon net-houses can be constructed for Rs. 6 to 8 lakhs. On an average, 5 laboures per day per acre are required for proper crop care and upkeep. In order to keep the maintenance cost as minimum as possible, areas with minimum extremities of weather condition such as Bangaluru, Pune, Nashik and surrounding places, are ideal for year round production of vegetables. As the cost of cooling is quite high, regions with hot summer conditions will add to the production costs, making the commercial vegetable production an economically viable proposition provided the market prices for the produce fetches higher premium price.

Conclusion

Although, greenhouse production is an intensive undertaking, it can be very satisfying and rewarding. One advantage of greenhouse vegetable production is that a relatively small area is required as compared with open field cultivation to achieve the similar target yield and produce with superior quality. In addition, the return on investment can be good if the requisite markets can be found.

Constraints in Protected Vegetable Cultivation

Greenhouse vegetable cultivation in India, though very ancient but still is in its infancy. Very little efforts have been made to exploit vast potentials of protected vegetable cultivation. Some major constraints and problems which restrict the vegetable production in greenhouse in our country are as follows:

1. Higher initial investment (construction and operational cost)
2. Unavailability of construction material, tools and implements at reasonable price timely
3. Lack of standardized designs of GH under different agro-climatic region of country
4. Lack of uninterrupted power supply and quality water sources
5. Inadequate research on development of suitable varieties and production technologies for protected cultivation
6. Lack of insurance coverage for structures and crop failure
7. Lack of govt. policies (providing incentives to beneficiaries, export policy) and bank loan facilities
8. Inadequate transport, storage and processing facilities of vegetables and flowers

Future Prospects and Strategies

Obviously, protected cultivation offers several advantages to grow high quality vegetables even under unfavorable climatic conditions. This technique requires high

input costs and good management. Overall, to make it successful venture growers need to make proper planning and vigil. Simultaneously, a policy framework to generate an inclusive growth for a country like India has to be consistent with the agro climatic and social conditions of the particular area. Thus for inclusive growth of this sector, emphasize must be given to the following aspects:

1. Development of location specific cost-effective design of GH and manufacturing
2. Standardization of package and practice including fertigation for different agro-climatic condition
3. Evolving effective and eco-friendly control measures of pathogens
4. Reducing dependency to import seeds and development of suitable varieties within country
5. Intensification of R&D on development of tools, devices, and equipments to reduce cost of production
6. Development of biodegradable polymers as cladding materials

References

Albright LD. 2002. Controlling greenhouse environments. *Acta Hort*, 578:47-54

Bansal N.K and Minke G.1988. Climatic zones and rural housing in India. Part I of the Indo-German Project of Passive Space Conditioning: 19-21.

Bisht, B.S.2012.Prospects and scenario of protected cultivation of horticultuural crops in Uttarakhand.- *Proceedings of* National seminar on protected cultivation of vegetables and flowers- A value chain approach held at G.B.P.U.A&T, Pantnagar. Uttarakhand.1-8 pp.

Castilla N.J, Hernandez A.F and Abou- Hadid. 2004. Strategic crop and greenhouse management in mild winter climate areas. *Acta Hort*, 486: 301-5.

Chadha, K.L.2001.Hi-Tech Horticulture in India.Policy Paper 13.national Academy of Agricultural Sciences, New Delhi.

Chadha, R.2008.High value agriculture in India. In *Souvenir*- 3rd Indian Horticulture congress. The Horticulture Society of India, New Delhi.

Garnaud, J. C. 1987. A survey of the development of plasticulture: Question to be answered. *Plasticulture* 74: 5-14

Janakiram.T. 2017. Employment and entrepreneurship opportunities in floriculture. In *Souvenir* Advances in Floriculture and Landscape Gardening (S.K. Malhotra and Lallan Ram Eds). pp. 41-46.Central Institute of Horticulture Department of Agriculture, Co-operation and Farmers Welfare. Ministry of Agriculture and Farmers Welfare.Government of India, Medziphema, Dimapur, Nagaland - 797106.

Jensen M.H. 2002. Controlled environment agriculture in deserts, tropics and temperate regions. A world review. *Acta Hort*, 578:19-25

Malhotra,S.K.2017.Emerging floriculture industry in India. In *Souvenir* Advances in Floriculture and Landscape Gardening(S.K. Malhotra and Lallan Ram Eds).pp 32-40.Central Institute of Horticulture Department of Agriculture, Co-operation and Farmers Welfare. Ministry of Agriculture and Farmers Welfare.Government of India, Medziphema, Dimapur, Nagaland - 797106.

Nelson P.V. 1978. Green House Operation and Management. Reston Publ.Co.

Rao,V.K.2013. Production system and growing media for greenhouse cultivation of flower crops. *In proceedings* of winter school on Hi-Tech Technologies for enhancing productivity, nutritional quality and value addition in fruits and flower productions. December, 10-30, 2013 at G.B.P.U.A&T, Pantnagar. Uttarakhand. pp. 317-326.

Reddy S, JanakiramT, Balaji T, Kulkarni S and Misra R.L. 2007.Hi-tech Floriculture. Indian Society of Ornamental Horticulture, New Delhi.

Sindhu S.S and Singh, B.2012.Research highlights in protected cultivation of flricultural crops. *Proceedings* National seminar on protected cultivation of vegetables and flowers- A value chain approach. G.B.P.U.A&T, Pantnagar. Uttarakhand. pp. 96-102.

Singh, Awani Kumar, ShriDhar, Singh, M.C. and Shilza Thakur.2013.Hi-Tech Technologies for enhancing productivity, nutritional quality and value addition in horticultural crops. -*Proceedings* of winter school on Hi-Tech Technologies for enhancing productivity, nutritional quality and value addition in fruits and flower productions. December, 10-30, 2013 at G.B.P.U.A&T, Pantnagar. Uttarakhand. 317-326 pp.

Singh, B.2012. Protected cultivation technology for diversification and livelihood security.*Proceedings* National seminar on protected cultivation of vegetables and flowers- A value chain approach.G.B.P.U.A&T, Pantnagar. Uttarakhand. pp. 27-32.

Singh, H.P. and Malhotra, S.K.2012.Current scenario and policy issue for protected cultivation of Horticultural crops in India- P*roceedings* of National seminar on protected cultivation of vegetables and flowers- A value chain approach.G.B.P.U.A&T, Pantnagar. Uttarakhand. pp. 9-20.

Singh,B.2014. Protected cultivation of Horticultural crops in India-Challenges and opportunities. Agrotechnology. 2(4), 1 p.

Sirjacobs M. 1988.Agro-climatological criteria for selecting the most appropriate areas for protected cultivation in Egypt. In: "Protected cultivation in the Mediterranean climate- Greenhouses in Egypt." F.A.O Rome, Italy: 5-12.

Tuzel Y, Gul A and Tuzel I.H 2004. Different soil-less culture systems. Regional Training Workshop on Soil-less Culture Technologies: 69-85. 3rd to 5th March, Izmir, Turkey.

Wittwer S.H and Castilla N.1995. Protected cultivation of horticultural crops, Worldwide. *Hort. Tech.* 5: 6-23.

Index

F

G

H

I

K

L

M

N

O

P

R

S

T

V